THE SMARTNESS MANDATE

THE SMARTNESS MANDATE

ORIT HALPERN AND ROBERT MITCHELL

THE MIT PRESS   CAMBRIDGE, MASSACHUSETTS   LONDON, ENGLAND

The MIT Press would like to thank the anonymous peer reviewers who provided comments on drafts of this book. The generous work of academic experts is essential for establishing the authority and quality of our publications. We acknowledge with gratitude the contributions of these otherwise uncredited readers.

This book was set in Stone Serif by Westchester Publishing Services, Danbury, CT. Printed and bound in the United States of America.

Library of Congress Cataloging-in-Publication Data

Names: Halpern, Orit, 1972– author. | Mitchell, Robert, 1969– author.
Title: The smartness mandate / Orit Halpern and Robert Mitchell.
Description: Cambridge, Massachusetts : The MIT Press, [2022] | Includes
  bibliographical references and index.
Identifiers: LCCN 2022000690 (print) | LCCN 2022000691 (ebook) |
  ISBN 9780262544511 (paperback) | ISBN 9780262371957 (epub)
Subjects: LCSH: Artificial intelligence—Industrial applications. | Artificial
  intelligence—Philosophy.
Classification: LCC TA347.A78 H34 2022 (print) | LCC TA347.A78 (ebook) |
  DDC 006.3—dc23/eng/20220528
LC record available at https://lccn.loc.gov/2022000690
LC ebook record available at https://lccn.loc.gov/2022000691

10 9 8 7 6 5 4 3 2 1

CONTENTS

ACKNOWLEDGMENTS

Orit Halpern would like to thank the Haus der Kulturen der Welt for support and for a forum for the early research. She also would like to thank the AUDACE program of Les Fonds de Recherche du Québec, the Graham Foundation, the Digital Now Architecture and Intersectionality Mellon Program at the Canadian Center for Architecture, and the Max Planck Institute for the History of Science in Berlin for their support for the research in this book. Particularly, Lorraine Daston and David Sepkoski at the Max Planck Institute offered a supportive environment and an intellectual milieux for developing these ideas. In Canada, Yuri Furuhata and Marc Steinberg were important interlocutors and inspirations for this text. Joshua Neves was critical in creating multiple opportunities and forums to workshop this research. Alessandra Ponte was a collaborator and teacher on all matters of landscape, environment, and industry 4.0. Nanna Bonde Thylstrup offered a residency, workshop opportunities, and inspiration for thinking about digital data in Copenhagen. Claudia Mareis, Kenny Cupers, and Johannes Bruder in Basel, Switzerland, all deserve mention for producing a space for rethinking of design, technology, and politics. Orit also would like to thank Ned Rossiter and Brett Neilsen for the many opportunities to work on logistical and big data infrastructures globally, and Jonathan Roberge for input and editorial assistance in developing the material on the Atacama and Chile. Ezekiel

Dixon-Román gave invaluable input concerning resilience and race, and Sudipto Basu provided research assistance in Kolkata. Finally, she could not have completed this work without the support and constant encouragement of her family: Tal, Atara, Mordechai, Iris, and Galia Halpern.

Robert Mitchell would like to thank the Haus der Kulturen der Welt for supporting an early version of some of these arguments, and the Hanse-Wissenschaftskolleg Institute for Advanced Study, Delmenhorst, Germany, for a fellowship that helped to make this book possible. He also is grateful to Inga, Kaia, and Nankea for their patience and encouragement.

We would like to thank the editors of *Grey Room* for their enthusiastic reception and publication of an early version of our introduction (*Grey Room* 68 [2017]: 106–129). In several of our case studies we draw on data previously published in the following: Orit Halpern, "Resilient Natures," *Social Text*, November 24, 2020, https://socialtextjournal.org/periscope_article /resilient-natures/; Orit Halpern, "Planetary Intelligence," in *The Cultural Life of Machine Learning: An Incursion into Critical AI Studies*, ed. Jonathan Roberge and Michael Castelle (London: Palgrave Macmillan, 2021), 227–256; and Orit Halpern, "Golden Futures," *Limn*, no. 10 (January 2019): 107–114, https://limn.it/articles/golden-futures/. We thank MIT Press for its support of this book, and the three anonymous press reviewers for their constructive, encouraging, and helpful comments.

PROLOGUE: WELCOME TO THE SMART PLANET

Event horizon
   "a point of no *return*"
   "*a boundary beyond which events* cannot affect an observer on the opposite side of it"[1]

On April 10, 2019, the first image of a black hole was presented to humanity (figure P.1). To produce this miracle, scientists and engineers from a team spanning the globe turned the earth itself into a vast sensor. They called this earth-as-sensor the Event Horizon Telescope (EHT). Only a dish the size of the planet itself would be sensitive enough to collect weak electromagnetic signals from more than 50 million light-years away and, through this activity, provide empirical evidence for one implication of Albert Einstein's general theory of relativity.

When the image was released, it circulated at literally the speed of light across that most human and social of networks, the internet. Comments online ranged from amazement to frustration that the black hole seemed to look just like we thought it might: "Awesome," "amazing," "mystical," and "capable of making humans fall in love" jockeyed with "anticlimactic," "Really?," and "It looks like the Eye of Sauron from *The Lord of the Rings*."[2] Perhaps, the latter commentators seemed to be suggesting, the visual output of the process of turning our entire planet into a sensing

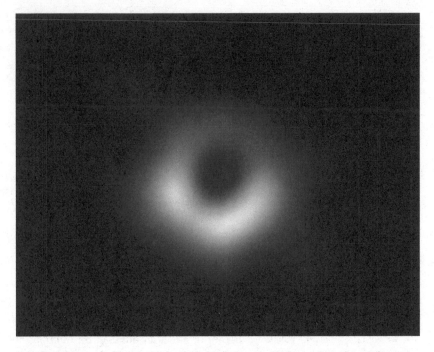

P.1 First image of a black hole, April 10, 2019. *Source*: NASA Jet Propulsion Laboratory, https://www.jpl.nasa.gov/edu/news/2019/4/19/how-scientists-captured-the-first-image -of-a-black-hole/.

technology was simply an artifact of computer graphics algorithms—that is, merely another stereotypical image that drew on and recalled long-standing Western cultural tropes of radically alien and powerful forces. By combining mythic and aesthetic conceptions of outer space and the power of the gods with the dream of scientific objectivity and a form of vision enabled by technology, the image of the black hole brought together two disjunctive temporalities: on the one hand, this event image crystallized a new imaginary of planetary (and even post-planetary) spaces fully integrated through data and machine sensing; on the other hand, the event image mobilized very old conventions of what extreme nonhuman alterity might look like, returning us to the legacies of myths and gods.

However one wants to understand the scientific truth of this image, we argue that the image itself provides evidence of a radical reformulation of perception and cognition. This image presents the figure of the terminal limits of human perception even as it also embodies a new form of experience and perception enabled by the literal networking of the entire planet into a sensor-perception instrument and experiment. This image is in this sense an allegory of the artificial intelligence and machine-learning systems that underpin it; it is both an image of a black hole and an image of an entire planet turned into a data-gathering machine. It is, we suggest, an image of *smartness*.

We open our prologue with this example as a way of pointing toward the basic question this book seeks to answer: namely, how have we come to see the planet and its denizens as data-collecting instruments? We suggest that this question requires us to rethink conventional approaches to smartness. The term "smartness" now usually comes attached to some digital technology or environment—for example, smartphones, smart cities, or smart cars. Critical analyses have understandably tended to focus on the specific genealogies, promises, and perils of those spaces or devices. We propose, however, that smartness, as the term itself suggests, should be understood as first and foremost an epistemology: that is, a way of knowing and representing the world so that one can act in and upon that world. This epistemology relies on new practices, technologies, and subjects, and these include artificial intelligence and machine learning, which are the bedrock of smartness. However, as we will note in the chapters that follow, treating smartness as an epistemology means approaching artificial intelligence and machine learning not simply as techniques for solving problems of calculation but rather as modes of addressing the world in terms of logic gates, networked intelligence, and distributed populations.

To clarify this point, we provide here several additional orienting images, each linked in different ways to the EHT image of the event horizon. Our second and third images document a key installation in the EHT, the Atacama Large Millimeter/Submillimeter Array (ALMA) installation (figures P.2 and P.3). On March 13, 2017, one of this book's authors (Halpern) visited this installation. Located on the Chajnantor Plateau in the Atacama Desert in Chile, the radio telescopes are positioned at an elevation of 5,050 meters in one of the driest and most extreme environments on earth.

If the event horizon is a point of no return, the Atacama is the landscape of that horizon, the infrastructure for our imaginaries of abandoning earth and leaving the past behind. NASA and other space agencies use this desert to test equipment, train astronauts, and study the possible

P.2 High-altitude submillimeter wave array, ALMA Observatory, Chajnantor Plateau, Atacama, Chile. Part of the EHT. *Source*: Photo by Orit Halpern, March 13, 2017.

astrobiology of the planets we will purportedly colonize in the future. There is, of course, something of an irony here, for the goal of ALMA itself is to collect history: every signal processed by the installation is eons old, having traveled millions or billions of light-years in time-space.

To produce the event horizon image, scientists used interferometry, a process that correlates the different radio waves captured by many telescopes into a singular representation. The key difficulty in this process is removing the massive amounts of noise in the data so that only those signals matching the theory of relativity's predictions for a black hole are correlated across EHT collection sites. Since a black hole (and, in fact, any stellar object) is very small compared to the scale of space, data from many different sources enters the receiving dishes, and so the signal-to-noise ratio is minuscule. Only machines have the capacity to analyze this quantity of data. However, machines can do this only when they are assisted by human "data cleaners." These data-cleaning teams employ different machine-learning approaches—including so-called unsupervised learning

P.3 ALMA towing vehicles, ALMA Observatory, Chajnantor Plateau, Atacama, Chile. Part of the EHT. *Source*: Photo by Orit Halpern, March 13, 2017.

methods, in which computers try out algorithms on their own—in an effort to identify artifacts in the data and remove them.[3] This was, however, quite difficult since no one had previously "seen" an event horizon so no one knew exactly what counted as signal and what counted as noise. In the face of such uncertainty, machines help us decide what is meaningful and what is not.

ALMA is one of the nodes of the network that enabled the EHT to turn the earth itself into a medium for capturing black hole data. However, the physical location of ALMA—namely, the Atacama Desert—allows us to understand several other ways in which earth itself has become a medium for smartness. Katie Detwiler, an anthropologist working in the Atacama (and Halpern's guide), notes that many here repeat the mantra that "Chile is copper." Copper is fundamental to smartness, for its electrical conductive properties mean that this element is in essentially every machine on earth, including ALMA. The Atacama Desert contains some of the largest copper mines on earth, and the excavation of copper from this area and the threading of this element throughout the world in the form of devices is another way in which the earth becomes a medium.

Moreover, copper mining itself has become smart, in part by turning the inside of the mine into a smart environment. At the Center for Mathematical Modeling at the University of Chile, located in Santiago, some 1,600 kilometers south of ALMA, Alejandro Jofré —one of the center's lead scientists in mathematical modeling and trained in optimization and game theory—explains that the center aims to bring the best in mathematical modeling to bear on questions of mine optimization, discovery, and supply chain management (figures P.4 and P.5). Reducing costs around, and improving, exploration is critical, as this is the most difficult and expensive part of the extraction industry process, often resulting in no return on investment. Better modeling and machine learning allows mine managers to extend their operations, discover ever more minute deposits of ore, and continue to expand extraction. And the application of artificial intelligence and big data solutions in geology and in mine management has in fact increased Chile's contribution to the global copper market: despite increasingly exhausted mine repositories, Chile increased its global share of copper for industrial use from 16 percent in 1990 to 30 percent in 2020.[4] And Codelco, the Chilean

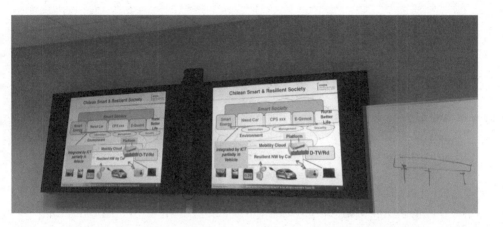

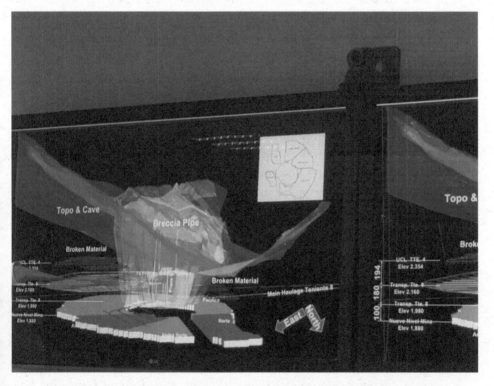

P.4 and P.5 Dr. Alejandro Jofré presenting on real-time analytics for decision-making in extraction. Center for Mathematical Modeling, University of Chile, Santiago. *Source*: Photos by Orit Halpern, March 21, 2017.

state-owned copper conglomerate, has entered major agreements with Uptake, a Chicago-based artificial intelligence and big data enterprise platform provider.[5]

A few miles from ALMA lies another landscape of extraction, metal, and energy, also linked to the stars and our future(s). SpaceX, Tesla, and the high-tech industries that in theory will eventually allow us to rid ourselves of the old heavy-industrial and carbon-based economies all bank on the Atacama, for this desert contains not only copper but also the "new gold," lithium (figure P.6). Lithium is the lightest of metals, and also one of the oldest (it was one of only three elements created at the time of the Big Bang). Lithium provides the key ingredient for the batteries that power most of our smart devices and is thus a central part of the anticipated greener futures of both machines and energy.

The salt beds from which lithium is extracted via brine are beautiful. Lithium is never pure but always mixed with other elements, many of which are also valuable, such as magnesium and potassium. As one looks over the extraction fields, one sees an array of colors, running from yellow to bright blue. The first fields are still full of potassium, which can serve as the key element for fertilizers; as the beds dry longer, they turn bluer and then more yellow. Finally, after almost a year, the bed dries and lithium salt emerges.[6] The salt is scraped from the bed, harvested, separated from trace boron and magnesium, and affixed with sodium carbonate for sale. Alejandro Bucher, the technical manager of the Sociedad Química y Minera (SQM) installation, takes us on a tour. SQM, he tells us, is environmentally excellent, as almost no chemicals are used in the process. The extraction of lithium is solar powered—the sun dehydrates the water and draws off the salts—and is thus a "pure" process. Or rather, almost pure, since the process also requires a significant amount of water, which is generally lacking in a desert. He assures us, however, that recent and future technical advances will optimize this problem and make the environment resilient, as more efficient water evaporation capture systems and desalinization plants reduce the impact of lithium extraction on the desert and on these brine waters, which are also sites of fragile ecosystems of shrimp, unique bacteria, and flamingos.[7]

Environmentalists do not necessarily agree with this prediction, and the general process of assessing environmental impact here has been critiqued

P.6 Salar de Atacama, Sociedad Química y Minera fields. *Source*: Photo by Orit Halpern, March 23, 2017.

as opaque.[8] While copper mines in Chile are still state controlled, SQM is a private company. SQM has been attacked for anti–trade union practices, and unions are fighting to label lithium a matter of national security so the state can better regulate the material. SQM is also part of the planetary systems of logistics around belt roads and resources. In 2018, the Chinese corporation Tianqi acquired a 24 percent share of SQM, essentially coming to dominate the corporation. While the Chilean government continues to monitor the situation and seeks to limit Chinese participation on the board of the corporation, the situation is still in flux.[9] These logistic games demand privatized water supplies, which in turn threaten indigenous villages in the area, increasing the precarity of populations already understood by many as disposable.[10]

As we noted, these salt beds serve as fragile infrastructures for unique ecosystems and communities that are threatened by the extraction process. Ironically, one of the groups threatened by lithium extraction is another group of scientists, astrobiologists, who study the bacteria in these brines in order to imagine our future on other planets. The bacteria that live in these salt flats have evolved in a manner seen almost nowhere else on earth, and the extreme conditions of this environment might thus offer clues about life on Mars, survival in space, and forms of life that could exist on other planets. We cannot expect to be alone in the universe, and these bacteria allow us to envision, via their novel metabolisms and capacity to live under pH conditions lethal to most other organisms, other possibilities for life. Hence, these astrobiologists argue, we cannot afford to destroy these salt beds in order to make lithium batteries, for these beds are an irreplaceable source of knowledge for our survival elsewhere and for our understanding of how we might terraform other planets.[11] What, then, to do when the future that these batteries make possible also seems to disappear through the process of extraction necessary for the power source?

THE SMARTNESS MANDATE

We opened with this topography of Chile, and especially the Atacama, because it defines the possibilities and problems of what we call in this book *the smartness mandate*. Our visual topography also allows us to reference, while at the same time frame, earlier efforts to understand how contemporary life is governed by capitalism and calculation. Chile played an important role in Naomi Klein's pathbreaking book *The Shock Doctrine*, published a decade ago. Klein described this book as a letter or message from the front lines of neoliberalism, and her tour de force provided us with a new vocabulary—and, equally important, a new tactical map, or topographical representation, or image—for understanding how contemporary forms of capitalism operate. Klein took around 30 years of history and discovered a pattern in the data, which she called *disaster capitalism*. This pattern linked the actions of psychiatrists who experimented with electrical shocks on patients in the 1950s to the torture and massacre of political dissidents and the violent reorganization of the economy in the name of structural readjustment in the 1970s. Chile played a key role in Klein's account, for it was here, under the dictator Augusto Pinochet, that many of these psychological and economic transformations were first linked in the context of national politics.

Today, the doctrine of shock has never appeared more pertinent, especially as we finish writing this book in the midst of the COVID-19 pandemic. In our current moment, market volatility, planetary-scale disease tracking, and human suffering appear to be "naturally" connected, and saving the stock market has often seemed to some to be more important than saving human lives. Yet even as COVID-19 shares much with pandemics in the past, never has the threat of disease and species-wide danger been shared so *synchronically* by means of media. In other words, computational and digital technologies mark this event in unprecedented ways. We track curves, analytics, and numbers and assume that big data will allow us to manage the plague that has arrived. Automated platforms and social networks deliver our goods, mediate our work and friendships, trade stocks, and maintain what was once called the *social*. Artificial intelligence and machine learning are also being deployed (or at least imagined) to predict future disease curves and to rapidly discover, test, and simulate the molecular structures and compounds that might serve as treatments or vaccines. This turn to computation as the infrastructure of salvation is unique, we argue, to our present. Since so few of these computational capacities are actualized, this says much about our future imaginaries of life on this planet (and, within this imaginary, on other planets in the future). This machine dream signifies a new phase in both

betting on and experimenting with possible futures through computational techniques.

Our current situation thus begs the question of what has and what has
not changed over the last 50 years—that is, since the emergence of disaster
capitalism in the 1970s. Klein focused primarily on how natural and political disasters are made commensurable through the logic of experiment and
through technology. For psychiatrists, the experiment was located in the
transformation of the psyche and its reprogramming, while for "the Chicago Boys"—that is, neoliberal economists such as George Stigler and Milton Friedman, based at the University of Chicago—entire economies were
positioned as test beds for economic theories that these markets could help
turn into realities via the mediation of dictators such as Pinochet. However,
we argue that, in our present moment, cities, financial markets, and the
earth itself have become experiments for computation, and this represents
a development that exceeds Klein's concepts of the shock doctrine and
disaster capitalism.

This prologue thus returns to the site originally mapped as the first
"experiment" in shock and economy, Chile, and seeks to underscore,
through a new set of images, the new coordinates of our present. We
hope through this means to change our perspective on what smartness
denotes and what it might make possible. This serves as our own letter of
care for the effort to make this "new nature" less normal, queerer, and
different.

INTRODUCTION

On November 6, 2008, still in the immediate aftermath of the worldwide economic crisis initiated by the US subprime mortgage market collapse, then-chairman of IBM Sam Palmisano delivered a speech at the Council on Foreign Relations in New York City. The council is one of the foremost think tanks in the United States, its membership composed of senior government officials, members of the intelligence community (including the CIA), business leaders, financiers, lawyers, and journalists. Yet Palmisano was not there to discuss the fate of the global economy. Rather, he introduced his corporation's vision of the future in a talk titled "A Smarter Planet." In glowing terms, Palmisano laid out a vision of fiber-optic cables, high-bandwidth infrastructure, seamless supply chain and logistical capacity, a clean environment, and eternal economic growth, all of which were to be the preconditions for a "smart" planet. IBM, he argued, would lead the globe to the next frontier, a network beyond social networks and mere Twitter chats. This future world would come into being through the integration of humans and machines into a seamless Internet of Things that would generate the data necessary for organizing production and labor, enhancing marketing, facilitating democracy and prosperity, and—perhaps most importantly—for enabling a mode of automated, and seemingly apolitical, decision-making that would guarantee the survival of the human species in the face of pressing environmental challenges. In Palmisano's talk, "smartness" referred to the interweaving of dynamic, emergent computational networks with the goal of producing a more resilient human species—that is, a species able to absorb and survive environmental, economic, and security crises by perpetually optimizing and adapting technologies.[1]

Palmisano's speech was notable less for its content, which to a large degree was an amalgamation of existing claims about increased bandwidth, complexity, and ecological salvation, than for the way in which its economic context and planetary terminology made explicit a hitherto tacit political promise that had attended the rise of smart technologies. Though IBM had capitalized for decades on terms associated with intelligence and thought—its earlier trademarked corporate slogan was "Think"—by 2008 the adjective "smart" was attached to many kinds of computer-mediated technologies and places, including phones, houses, cars, classrooms, bombs, chips, and cities. Palmisano's "smarter planet" tagline drew on these earlier invocations of smartness, especially the notion that smartness required an extended infrastructure that produced an environment able to automate many human processes and respond in real time to human choices. His speech also underscored that smartness demanded an ongoing penetration of computing into infrastructure to mediate daily perceptions of life. (Smartphones, for example, are part of a discourse in which the world is imagined as networked, interactive, and constantly accessible through technological interfaces, and a smartphone's touch screen is in fact enabled by an infrastructure of satellite networks, server farms, and cellular towers, among many other structures that facilitate regular access to services, goods, and spatial location data.) But as Palmisano's speech made clear, these infrastructures now demanded an *infrastructural imaginary*—an orienting telos about what smartness is and does. This imaginary redefined no less than the relationships among technology, human sense perception, and cognition. With this extension of smartness to both the planet and the mind, what had been a corporate tagline became a governing project able to individuate a citizen and produce a global polity.

This new vision of smartness is inextricably tied to the language of crisis, whether the latter is a financial, ecological, or security event. But where others might see the growing precariousness of human populations as best countered by conscious planning and regulation, advocates of smartness instead see opportunities to decentralize agency and intelligence by distributing it among objects, networks, and life-forms. They predict that environmentally extended smartness will take the place of deliberative planning, allowing resilience in a perpetual transforming world. Palmisano proposed "infus[ing] intelligence into decision making" itself.[2] What Palmisano presented in 2008 as the mandate of a single corporation is in fact central to contemporary design and engineering thinking more generally.

We call these promises about computation, complexity, integration, ecology, and crisis *the smartness mandate*. We use this phrase to mark the fact that the assumptions and goals of smart technologies are widely accepted in global polity discussions and that they have encouraged the creation of novel infrastructures that organize environmental policy, energy policy, supply chains, the distribution of food and medicine, finance, and security policies. The smartness mandate draws on multiple and intersecting discourses, including ecology, evolutionary biology, computer science, and economics. Binding and bridging these discourses are technologies, instruments, apparatuses, processes, and architectures. These experimental networks of responsive machines, computer mainframes, political bodies, sensing devices, and spatial zones lend durable and material form to smartness, often allowing for its expansion and innovation with relative autonomy from its designers and champions.

This book critically illuminates some of the key ways in which the history and logic of the smartness mandate have become dynamically embedded in the objects and operations of everyday life—particularly the everyday lives of those living in the wealthier Global North but, for the advocates of smartness, ideally the lives of every inhabitant of the globe. This approach allows us to consider questions such as the following: What kinds of assumptions link the "predictive" product suggestions made to a global public by retailers such as Amazon or Netflix with the efforts of Korean urban-planning firms and Indian economic policy-makers to monitor and adapt in real time to the activities of their urban citizenry? What kinds of ambitions permit the migration of statistically based modeling techniques from relatively banal consumer applications to regional and transnational strategies of governance? How do smart technologies that enable socially networked applications for smartphones—for example, the Microsoft Teams app, which enables distributed multisite and multiuser conversation and workflow and is used by 75 million registered users a day (located primarily in the US, Europe, Latin America, and Asia)—also cultivate new forms of global labor and governmentality, the unity of which resides in the coordination via smart platforms rather than, for example, geographical proximity or class?[3] Each of these examples relies upon the mediation of networks and technologies that are designated to be smart, yet the impetus for innovation and the agents of this smartness often remain obscure.

We see what is still the relatively short history of smartness as a decisive moment in histories of reason and rationality. In their helpful account of what they call "Cold War rationality," Paul Erickson and his colleagues have argued that in the years following World War II, American science, politics, and industry witnessed "the expansion of the domain of rationality at the expense of . . . reason," as machinic systems and algorithmic procedures displaced judgment and discretion as ideals of governing rationally.[4] Yet at the dawn of the twenty-first century, Cold War rationality gave way to the tyranny of smartness, an eternally emergent program of real-time, short-term calculation that substitutes *demos* (i.e., provisional models) and simulations for those systems of artificial intelligence and professional expertise and calculation imagined by Cold War rationalists. In place of Cold War systems based on "rational" processes that could still fall under the control and surveillance of centralized authorities or states, the smartness mandate embraces the ideal of an infinite range of experimental existences, all based on real-time adaptive exchanges among users, environments, and machines. Neither reason nor rationality is understood as a necessary guide for these exchanges, for smartness is presented as a self-regulating process of *optimization* and *resilience* (terms that, as we note below, are themselves moving targets in a recursive system).

Whereas Cold War rationality was highly suspicious of innovation, the latter is part of the essence of smartness. In place of the self-stabilizing systems and homeostasis that were the orienting ideal of Cold War theorists, smartness assumes perpetual growth and unlimited turmoil; destruction, crisis, and the absence of architectonic order or rationality are the conditions for the possibility for smart growth and optimization. Equally important, whereas Cold War rationality emanated primarily from the conceptual publications of a handful of well-funded think tanks, which tended to understand national populations and everyday culture as masses that need to be guided, smartness pervades cell phones, delivery trucks, and health-care systems and relies on the interactions among, and the individual idiosyncrasies of, millions or even billions of individuals around the planet. Moreover, whereas Cold War rationality was dominated by the thought of the doppelgänger rival (e.g., the US vs. the USSR, the East vs. the West), smartness is not limited to binaries.[5] Rather, it understands threats as emerging from an environment that, because it is always more complex than the systems it encompasses, can never be captured in the simple schemas of rivalry or game theory. This in turn allows smartness to take on an ecological dimension: the key crisis is no longer simply that emerging from rival political powers or nuclear disaster but rather, more fundamentally, intrinsically unforeseeable events that will necessarily continue to emerge from an always too-complex environment.

If smartness is what follows after Cold War understandings of reason and rationality, the smartness mandate is the political imperative that smartness be extended to all areas of life. In this sense, the smart mandate is what comes after the *shock doctrine*, powerfully described by Naomi Klein and others.[6] As Klein notes in her book of the same name, the shock doctrine was a set of neoliberal assumptions and techniques that taught policy-makers in the 1970s to take advantage of crises to downsize government and deregulate in order to extend the rationality of the free market to as many areas of life as possible. The smart mandate, we suggest, is the current instantiation of a new technical logic with equally transformative effects on conceptions and practices of governance, markets, democracy, and even life itself. Yet where the shock doctrine imagined a cadre of experts and advisers deployed to various national polities to liberate markets and free up resources during moments of crisis, the smartness mandate both understands crisis as the normal human condition and extends itself by means of a field of plural agents—including environments, machines, populations, and data sets—that interact in complex manners and without recourse to what was earlier understood as reason or intelligence. If the shock doctrine promoted the idea that systems had to be fixed so that natural economic relationships could express themselves, the smartness mandate aims instead at resilience and practices management without ideals of futurity or clear measures of success or failure. We

describe this imperative of developing and instantiating smartness everywhere as a *mandate* in order to capture both its political implications—
though smartness is presented by its advocates as politically agnostic, it
is more accurate to see it as reconfiguring completely the realm of the
political—and the premise that smartness is only possible by drawing upon
the *collective intelligence* of large populations.

We illuminate the deep logic of smartness and its mandate in four
chapters, each focused on a different aspect of the smartness mandate.
These chapters take up the following questions:

1. What is the *agent* of smartness (i.e., what, precisely, enacts or possesses
   smartness)?
2. Where does smartness happen (i.e., what kind of *space* does smartness
   require)?
3. What is the key operation of smartness (i.e., what does smartness *do*)?
4. What is the purported result of smartness (i.e., at what does it *aim*)?

Our answers to these four questions are as follows:

1. The (quasi-)agents of smartness are *populations*.
2. The territory of smartness is *the experimental zone*.
3. The key operation of smartness is *derivation*.
4. Smartness produces *resilience*.

Focusing on how the logics and practices of populations, experimental
zones, derivation, and resilience are coupled enables us to illuminate not
simply particular instantiations of smartness—for example, smart cities,
grids, or phones—but smartness more generally and its mandate ("every
process must become smart!").

Our analysis draws inspiration from Michel Foucault's concepts of governmentality and biopolitics, Gilles Deleuze's brief account of "the control society," and critical work on immaterial labor. We describe smartness
genealogically—that is, as a concept and practices that emerged from the
coupling of logics and techniques from multiple fields, including ecology, computer science, and government policy. We also link smartness to
the central object of biopolitics—namely, populations—and see smartness
as bound up with the key goals of biopolitics and governmentality. We
emphasize the importance of a mode of control based on what Deleuze
describes as open-ended modulation, rather than the permanent molding

of discipline. We also underscore the centrality of data drawn from the everyday activities of large numbers of people. Yet insofar as smartness positions the global environment as the fundamental orienting point for all governance—that is, as the realm of governance that demands that all other problems be seen from the perspective of experimental zones, populations, resilience, and optimization—the tools offered by existing concepts of biopolitics, the control society, and immaterial labor take us only part of the way in our account.[7]

POPULATIONS

Populations are the agents—or more accurately, the *enabling medium*—of smartness. Smartness is located neither in the source (producer) nor in the destination (consumer) for a product such as a smartphone but is rather the outcome of the algorithmic manipulation of billions of traces left by thousands, millions, or billions of individual users. Smartness requires these large populations, for they are the medium of what we will call the *partial perceptions* within which smartness emerges. Although, as we discuss below, these populations should be understood as fundamentally biopolitical in nature, it is more helpful first to recognize the extent to which smartness relies on an understanding of populations drawn from twentieth-century biological sciences such as evolutionary biology and ecology (figure I.1).

Biologists and ecologists often use the term "population" to describe large collections of individuals with the following characteristics:

1. Each member of the population differs at least slightly from one another.
2. These differences allow some individuals to be more "successful" vis-à-vis their environment than other individuals.
3. There is a form of memory that enables differences that are successful to appear again in subsequent generations.
4. As a consequence of (3), the distribution of differences across the population tends to change over time.[8]

This emphasis on the importance of individual differences for long-term fitness thus distinguishes this use of the term "population" from more common political uses of the term to describe the individuals who live within a political territory.[9]

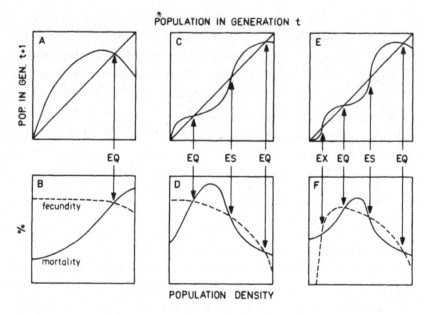

I.1 Diagram speculating on various futures for population reproduction curves and deriving fecundity and morbidity (*bottom row*) from these curves. *Source*: C. S. Holling, "Resilience and Stability of Ecological Systems," *Annual Review of Ecological Systems* 4 (1973): 1–23, 21.

Smartness takes up a biologically oriented concept of population but repurposes it for nonbiological contexts. Smartness presumes that each individual is not only biologically distinct but also distinct in terms of "social" characteristics such as habits, knowledge, and consumer preferences, and that information about these individual differences can be usefully grouped together so that algorithms can locate subgroupings of this data that thrive or falter in the face of specific changes. Though the populations of data drawn from individuals may map onto traditional biological or political divisions, groupings and subgroupings might also revolve around consumer preferences and could be drawn from individuals in widely separated geographical regions and polities (for example, Netflix's populations of movie preferences are currently created from users distributed throughout 190 countries).[10] Moreover, though these data populations are (generally) drawn from humans, they are best understood as distinct from the human populations from which they emerge: these are simply data populations of, for example, preferences, reactions, or abilities.

This is true even in the case of information drawn from human bodies located in the same physical space. In the case of the smart city, the information streaming from Fitbits, smartphones, credit cards, and transport cards is generated by human bodies in close physical proximity to one another, but individual data populations are then agglomerated at different temporalities and scales, depending on the problem being considered (for example, transportation routing, energy use, or consumer preferences). These discrete data populations enable processes to be optimized (i.e., enable "fitness" to be determined), which in turn produces new populations of data and hence a new series of potentialities for what a population is and what potentials these populations can generate.

A key premise of smartness is that while each member of a population is unique, the population is also "dumb"—that is, limited in its perception—and that smartness emerges as a property of the population as a whole only when these limited perspectives are linked via environment-like infrastructures. Returning to the example of the smartphone operating in a smart city, the phone becomes a mechanism for creating data populations that operate without the cognition or even the direct command of the subject (the smartphone, for example, automatically transmits its location and can also transmit other information about how it has been used). If populations enable long-term species survival in the biological domain, then populations enable smartness in the cultural domain, provided that populations are networked together with smart infrastructures. Populations are part of the perceptual substrate that enables modulating interactions among agents within a system that sustains particular activities. The infrastructures ensure, for example, that "given enough eyeballs, all bugs are shallow" (Linus's law); that problems can be crowdsourced; that there can be collective intelligence; and so on.[11]

This creation and analysis of data populations is clearly biopolitical in the sense initially outlined by Michel Foucault, but it is also vital to recognize smartness as a significant mutation in the operation of biopolitics. As Foucault stressed, the concept of population was central to the emergence of biopolitics in the late eighteenth century, for it denoted a collective body that had its own internal dynamics (of births, deaths, illness, etc.), which were quasi-autonomous in the sense that they could not be commanded or completely prevented by legal structures but could nevertheless be subtly altered through biopolitical regulatory techniques and technologies (for example, required inoculations or free-market mechanisms).[12] On the one hand, smartness is biopolitical in this same sense, for the members of its populations—movie watchers, cell phone users, health-care purchasers and users, and so on—are assumed to have their own internal dynamics and regularities, and the goal of gathering information about these dynamics is not to discipline individuals into specific behaviors but rather to find points of leverage within these regularities that can produce more subtle and widespread changes.

On the other hand, the biopolitical dimension of smartness cannot be understood as simply "more of the same" for four reasons. First, and in keeping with Deleuze's reflections on the control society, the institutions that gather data about populations are now more likely to be corporations rather than the state.[13] Second, and as a consequence of the first point, smartness's data populations often concern not those clearly biological events on which Foucault focused but rather variables such as attention, consumer choices, and transportation choices. Third, although the data populations that are the medium of smartness are drawn from populations of humans, this data relates differently to individuals than do Foucault's more health-oriented examples. Data populations themselves often do not need to be (and sometimes cannot be) mapped directly back onto discrete human populations: one is often less interested in discrete events that happen only once in the individual biographies of the members of a polity (e.g., a smallpox infection) than in frequent events that may happen across widely dispersed groups of people (e.g., movie preferences). The analysis of these data populations is then used to create, via smart technologies, an individual and customized information environment around each individual, which aims not to discipline individuals, in Foucault's sense, but to extend ever deeper and further the quasi-autonomous dynamics of populations. Fourth, in the case of systems such as high-speed financial trading and derivatives and in the logistical management of automated supply chains, entire data populations are produced and acted upon directly through machine-to-machine data gathering, communication, analytics, and action.[14] These new forms of automation and of producing populations mark transformations in both the scale and intensity of the interweaving of algorithmic calculation and life.

ZONES

Smartness has to happen somewhere. However, because the agents, or media, of smartness are globally distributed populations, the geography of smartness no longer follows traditional political borders. Advocates of smartness generally imply or explicitly note that the space of smartness is not that of the national territory. Palmisano's invocation of a smarter planet, for example, emphasizes the extraterritorial space that smartness requires: precisely because smartness aims in part at ecological salvation, its operations cannot be restricted to the limited laws, territory, or populations of a given national polity. Designers of smart homes likewise imagine a domestic space freed by intelligent networks from the physical constraints of the traditional home while the fitness app on a smartphone conditions the training of a single user's body through iterative calculations correlated with thousands or millions of other users across multiple continents.[15] These activities all occur in space, but the nation-state is neither their obvious nor necessary container, nor is the human body and its related psychological subject their primary focus, target, or even paradigm (e.g., smartness often employs entities such as *swarms* that are never intended to cohere in the manner of a rational or liberal subject).

At the same time, smartness also depends on complicated and often delicate infrastructures, such as fiber-optic cable networks and communication systems capable of accessing satellite data or server farms that must be maintained at precise temperatures or safe shipping routes that are invariably located at least in part within national territories and often subsidized by federal governments. Smartness thus also requires the support of legal systems and policing to protect and maintain these infrastructures, and most of the latter are provided by nation-states (even if partially in the form of subcontracted private security services).[16]

This paradoxical relationship of smartness to national territories is best understood as a mutation of the contemporary form of space known as zones. Related to histories of urban planning and development, where zoning has long been an instrument in organizing space, contemporary zones have new properties married to the financial and logistical practices that underpin their global proliferation. In the past two decades, urban historians and media theorists have redefined the zone in terms of its connection to computation and described the zone as the dominant territorial configuration of the present. As architectural theorist Keller Easterling notes, the zone should be understood as a method of *extrastatecraft* intended to serve as a platform for the operation of a new "software" for governing human activity. Brett Nielsen and Ned Rossiter invoke the figure of the *logistical city* or zone to make the same point about governmentality and computation.[17]

Zones do not denote the demise of the state but rather the production of new forms of territory. One important modality of this new form is a space of exception to national and often international law. A key example is the so-called free-trade zone. Free-trade zones are a growing phenomenon, stretching from the Pudong district in Shanghai to the Cayman Islands to the business districts and port facilities of New York State, and are promoted as conduits for the smooth transfer of capital, labor, and technology globally (with "smooth" defined as a minimum of delay as national borders are crossed). Free-trade zones are in one sense discrete physical spaces, but they also require new networked infrastructures linked through the algorithms that underwrite geographic information systems, global positioning systems, and computerized supply chain management systems, as well as the standardization of container and shipping architecture and regulatory legal exceptions (to mention just some of the protocols that produce these spaces). Equally as important is that zones are understood to be outside the legal structure of a national territory, even if they technically lie within its space.[18]

In using the term "zone" to describe the space of smartness, our point is not that smartness happens in places such as free-trade zones but rather that smartness aims both to globalize and, simultaneously, render more experimental the logic of zones. This logic of geographic abstraction, detachment, and exemption is exemplified even in a mundane consumer item such as activity monitors—for example, the Fitbit—that link data about the physical activities of a user in one jurisdiction with the data of users in other jurisdictions. This logic of abstraction is more fully exemplified by the emergence of so-called smart cities. An organizing principle of the smart city is that automated and ubiquitous data collection will drive, and perhaps replace, civic governance and public taxation. This ideal of a "sensorial" city that serves as a conduit for data gathering and circulation is a primary fantasy enabling smart cities, grids, and networks. Consider, for example, a prototype *greenfield*—that is, designed from the ground up—smart city development, such as Songdo in South Korea (figure I.2). This smart city is designed with a massive sensor infrastructure for collecting traffic, environmental, and closed-circuit television data and includes individual smart homes (apartments) with multiple monitors and touch screens for temperature control, entertainment, lighting, and cooking functions. The city's developers also hope that these living spaces will eventually monitor multiple health conditions through home testing. Implementing this business plan, however, will require significant changes to, or exemptions from, South Korean laws about transferring health information outside of hospitals. Lobbying efforts for this juridical change have been promoted by Cisco Systems (a US-based network infrastructure provider), the Incheon Free Economic Zone (the governing local authority), and POSCO (a Korean chaebol, i.e., a large, family-controlled

I.2   Ideal zonal imaginaries for cities. *Left*: Utopia by Sir Thomas More (1518). *Upper middle*: Sforzinda by Filarete (fifteenth–sixteenth century). *Lower middle*: Coevorden (the Netherlands, early seventeenth century). *Upper right*: Jurong Island, Singapore. *Lower right*: Songdo smart city, Incheon, South Korea, master plan, 2012. *Source*: Orit Halpern, Jesse LeCavalier, and Nerea Calvillo, "Test-Bed Urbanism," *Public Culture* 25, no. 2 (March 2013): 272–306, 275.

conglomerate that, in this case, focuses on construction and steel refining), formerly the three most dominant forces behind Songdo.

What makes smart territories unique in a world of zonal territories is the specific mode by which smartness colonizes space through the management of time (and this mode also helps explain why smartness is so successful at promulgating itself globally). We focus in this book on the prototype or demo nature of contemporary zones and the relationship of prototyping to catastrophe. As underscored in our opening example of Palmisano's speech, smartness is predicated on an imaginary of crisis that is to be managed through a massive increase in sensing devices, which in turn purportedly enable self-organization and constant self-modulating and self-updating systems. That is, smart platforms link zones to crisis via two key operations: a temporal operation, by means of which uncertainty about the future is managed through constant redescription of the present as a version, demo, or prototype of the future, and an operation of self-organization, through which earlier discourses about structures and *the social* are replaced by concerns about infrastructure, a focus on sensor systems, and a fetish for big data and analytics, which purportedly can direct development even in the absence of clearly defined ends or goals.

To put this another way, so-called smart cities such as Songdo follow a logic of software development. Every present state of the smart city is understood as a demo or prototype of a future smart city; every operation in the smart city is understood in terms of testing and updating. Engineers interviewed at Songdo openly spoke of it as an "experiment" and as a "test," admitting that many parts of the system currently did not work but stressing that problems could be fixed in the next instantiation elsewhere in the world.[19] As a consequence, there is never a finished product but rather infinitely replicable yet always preliminary, never-to-be-completed versions of these cities around the globe.

This temporal operation of demo-ing is linked to an ideal of self-organization. Smartness largely refers to computationally and digitally managed systems, from electrical grids to building management systems, that can learn and, in theory, adapt by analyzing data about themselves. Self-organization is thus linked to the operation of optimization (which we discuss in more detail below). Systems are to correct themselves automatically in relationship to their own operations. This organization is imagined as being immanent to the physical and informational system at hand—that is, optimized by computationally collected data rather than by external political or social actors. At the heart of the smartness mandate is thus a logic of immanence, by means of which sensor instrumentation adjoined to emerging and often automated methods for the analysis of large data sets allows a dynamic system to detect and direct its perpetual evolution.[20]

Our notion of zonal territories thus refers to a form of governance that is both spatial and temporal. The form of space is one of processes and practices, and we focus on the modulatory nature of these spaces. Smart zones are malleable: they are not static spaces, nor are they clearly delineated or taxonomically organized areas. Unlike the historic zoning of cities into commercial, private, and industrial spaces, demo-zones constantly rearrange these terms according to the mandates of emergency and computation. Instead of urban master plans or even utopian visions of cities that characterized (even if as ideals rather than as actual realities) earlier twentieth-century understandings of planning, smart zones operate instead by means of concepts of constant experimentation and feedback that transform space.

One of the key, and troubling, consequences of demo-ing and self-organization as the two zonal operations of smartness is that the overarching concept of crisis comes to obscure differences in various types of catastrophes. While every crisis event—for example, the 2008 subprime mortgage collapse or the Tohoku earthquake of 2011—is different, within the demo-logic that underwrites the production of smart and resilient cities, these differences can be subsumed under the general concept of crisis and addressed through the same methods (the implications of which must never be fully engaged because we are always demo-ing or testing solutions, never actually solving a stable underlying problem). Whether threatened by terrorism, subprime mortgages, energy shortages, or hurricanes, smartness always responds in essentially the same way. The demo is thus a form of temporal management that through its practices and discourses evacuates historical and contextual specificity of particular catastrophes and evades ever having to assess or represent the impact of these infrastructures because no project is ever "finished." It is this evacuation of differences, temporalities, and societal structures that most concerns us in confronting the extraordinary rise of ubiquitous computing and high-tech infrastructures as solutions to political, social, environmental, and historical problems.

DERIVATION

Smartness emerges when zones and increasingly fine-grained observations of the quasi-autonomous dynamics of populations are linked through *optimization processes* that are themselves oriented toward what we call a *logic of derivation*—that is, temporal technologies able to exploit current computational limits as both a present source of value and a hedge against an always unknowable and threatening future. Though optimization and derivation are quite different concepts and technical methods of optimization and derivation have different lineages, smartness links these two operations by orienting optimization toward the logic of derivation and derivatives.

Optimization as a concept and set of techniques is often understood as a synonym for "efficiency" and is equated with the techniques of industrial production and the sciences of efficiency and fatigue pioneered in the late nineteenth and early twentieth centuries by Fredrick Winslow Taylor and Frank Gilbreth.[21] In the context of smartness, though, notions of optimization defer and often displace older concerns with energy and entropy in important ways that separate its current reality from histories of efficiency. Though the history of optimization has yet to be written, the term itself seems to have entered common usage in English only in the 1950s via interrelated fields such as electrical engineering, computer research, and game theory.[22] For these discourses, "to optimize" meant to find the best relationship between the minima and maxima performances within a well-defined system or space. Optimization was not a normative or absolute measure of performance but an internally referential and relative one: for *this* system, given *these* goals and *these* constraints, the optimal solution is X. The effort to locate the one choice that provides the least cost and most benefit also defines much of the thinking about the economic agent in the second half of the twentieth century. The claim advanced by neoliberal economists beginning in the 1950s that *every* kind of conscious human activity, including choices about education, voting, and marriage partners, should be understood as fundamentally economic is also indebted to this understanding of technical optimization. Optimization is in this sense a key technique by which smartness promulgates the belief that everything—every kind of relationship among humans, their technologies, and the environments in which they live—can and should be algorithmically managed. Shopping, dating, exercising, the practice of science, the distribution of resources to public schools, the fight against terrorism, the calculation of carbon offsets and credits: all of these processes can—and must!—be optimized. It is in part this pursuit of "the best"—the fastest route between two points, the most reliable prediction of a product consumers will like, the least expenditure of energy in a home, the lowest risk and highest return in a financial portfolio, and so on—that implicitly justifies the term "smartness."

At the same time, however, twentieth-century technical optimization procedures are almost always linked to limits, or even failure, and smartness involves a very specific approach to such optimization failures—namely, deriving value from failure by means of "learning." This constitutes a break from the older models of efficiency grounded in energy consumption and materials. The development of calculus in the eighteenth century encouraged the hope that if one could simply find an equation for a curve that described a system, it would then always be possible in principle to locate the *absolute*, rather than simply local, maxima and minima for a system. Yet the problems engaged by twentieth-century electrical and computer engineers often had so many variables and dimensions that it was impossible, even in principle, to solve an equation completely. As computer scientist Dan Simon notes, even a problem as apparently simple as determining the most optimal route for a salesperson who needs to visit 50 cities would be impossible were one to try to calculate all possible solutions. There are $49!$ $(=6.1 \times 10^{62})$ possible solutions to this problem, which is beyond the capability of contemporary computing: even if one had a trillion computers, each capable of calculating a trillion solutions per second, and these computers had been calculating since the universe began—a total computation time of 15 billion years—they would not yet have come close to calculating all possible routes.[23]

In the face of the impossibility of determining the absolute maxima or minima for these systems using the so-called brute force approach (i.e., calculating and comparing all possible solutions), optimization often

involves finding good-enough solutions: maxima and minima that may
or may not be absolute but are more likely than other solutions to be
close enough to absolute maxima or minima to allow systems to continue operating without additional investment. The optimizing engineer
selects among different algorithmic methods that each produce, in different ways and with different results, good-enough solutions.[24]

Yet for real-world problems, any particular optimization method may
fail, in the sense that it becomes trapped by local minima or maxima (see
figure I.3). Smartness relies on seeing failed optimization as an occasion
for learning. In some cases, such learning is intended to mimic natural

I.3   The Ackley function for two-dimensional space: $f(x,y) = -20\exp\left[-0.2\sqrt{0.5(x^2+y^2)}\right]$
$-\exp[0.5(\cos 2\pi x + \cos 2\pi y)] + e + 20$. The absolute minimum of this function is zero.
However, since it contains many closely clustered local minima, some evolutionary optimization algorithms find the absolute minimum difficult to locate. Different evolutionary
optimization algorithms can thus be tested on this function to determine how close each
can get to the absolute minimum.

processes, especially computational ideals of biological evolution, which reframe local failure as part of a broader strategy of perpetual testing for new solutions. Evolutionary optimization algorithms, for example, begin with the premise that natural biological evolution automatically solves optimization problems by means of natural biological populations. It then seeks to simulate that process by creating populations of candidate solutions, which are mixed with one another (elements of one candidate solution are combined with elements of other candidate solutions) and culled through successive generations to produce increasingly good solutions. David B. Fogel, a consultant for the informatics firm Natural Selection Inc., which applies computational models to the streamlining of commercial activities, captures this sense of optimization as simply a continuation of nature's work: "Natural evolution is a population-based optimization process. Simulating this process on a computer results in stochastic optimization techniques that can often outperform classical methods of optimization when applied to difficult real-world problems."[25]

Optimization research implements these features (reproduction, mutation, competition, and selection) in computers in an effort to find "natural" laws that can govern the organization of industrial or other processes that, when implemented on a broad scale, become the conditions of life itself. The premise that systems can never be fully and finally optimized, if only because their environments change, also propels the demand for ever more sensors—more sites of data collection, whether via mobile device apps, hospital clinic databases, tracking of website clicks, and so on—so that optimization's realm can be perpetually expanded and optimization itself further optimized.

Yet treating failed optimization as an occasion for learning also requires time-based strategies for mitigating the consequences of such failures, and in the case of smartness, this means enframing optimization within a logic of derivation. A financial derivative—for example, a currency future option that gives the purchaser the right, but not obligation, to purchase the currency in the future at an exchange price agreed upon in the present—can be used to guard against the risk that the value of that underlying asset (the specific currency in question) will decrease in the future. These kinds of financial derivatives have been used by corporations that are based in one country but do business in another since significant changes in currency exchange rates can spell disaster for a company's bottom line if, for example, it invested in equipment at one currency exchange rate but several years later received payments for its products at a much lower currency exchange rate. Corporations and individuals can hedge against risk even more by bundling different derivatives together, resulting in derivatives that can eventually have very attenuated relationships to the underlying assets. (This was the case in the famous credit-swap derivatives that propelled the US housing market crisis that began in 2007, in which a single derivative might contain tiny slices of thousands of housing loans). As we will discuss at more length in our chapter on derivation, this operation is a means for managing uncertainty and for making what might otherwise be seen as extraordinarily dangerous or life-threatening decisions—for example, continuing to burn massive amounts of carbon despite clear evidence that this is changing the global climate for the worse—seemingly risk-free. Derivation thus enframes optimization by extracting value from the assumed repeat failure of optimization in the present and the demand to learn in the future. That is to say, derivative practices are betting that the future is not known, and the present may be imperfectly optimized—and this difference can be a source of speculation (see figure I.4).

The logic of derivation is perhaps most clearly exemplified by, but is not limited to, financial derivatives. We can see the same logic of reallocating risks (often unfairly) and deferring issues of responsibility in the arena of national security data analysis. Ethicist Louise Amoore describes how this same logic plays out in British homeland security software design. Software designers seek to help automate *risk flags* for border agents. While some of these risk flags are determined by traditional pieces of information that bear upon a traveler's identity—for example, passport or visa information—other details bear upon choices that do not seem intrinsic to personal identity at all, such as how close to departure a ticket was purchased, by what means it was purchased (cash or credit), and what meal a passenger selected. These latter pieces of information help establish an ever-changing norm of what "normal" travel looks like and allow the software program to compare each traveler with that shifting norm. Moreover, each time the software creates an erroneous red flag, that failure can be used to further refine the algorithm. Amoore calls this

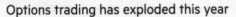

Options trading has exploded this year

Number of put and call option contracts traded in the US each day (m)

— Calls   — Puts

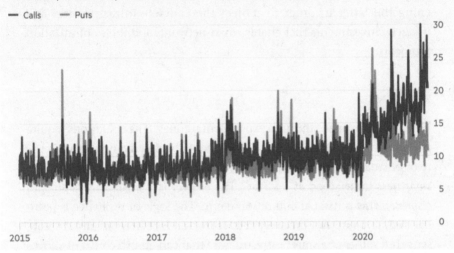

I.4   Chart of options trading during the COVID-19 pandemic. *Source: Financial Times,* December 21, 2020, https://www.ft.com/content/19cb6aa3-a390-4ed6-a695-9a1e70 0f35b6.

separation of data from the individual and its rebundling with thousands of other pieces of data, all with an eye toward determining whether an individual might pose a security risk, a *data derivative.*[26] In chapter 3, on derivation, we take up her point that, through such automation, responsibility for decisions is deferred or evaded, just as financial derivatives allow traders to hedge against risk without becoming legally responsible for the shaky investments they enable. That is, in derivative logic, value is extracted by shorting the bet, which also means never having to engage the consequences of an action or the future produced through these trades.

The derivative logic of optimization serves to justify the extension and intensification of the zonal logic of smartness. In order to optimize all aspects of existence, smartness must be able to locate its relevant populations (of preferences, events, etc.) wherever they occur. However, this is only possible when every potential data point (i.e., partial perception) on the globe can be directly linked to every other potential data point without interference from specific geographic jurisdictional regimes. As we noted above, this does not mean the withering of geographically based security apparatuses; on the contrary, derivation often requires strengthening the latter in order to protect the concrete infrastructures and financial investments that enable smart networks and their optimization protocols.

RESILIENCE

If smartness happens through experimental zones, if its operations require populations, and if it aims most fundamentally at derivation, what is the telos of smartness itself—that is, at what does smartness aim, and why is smartness understood as a virtue? The answer is that smartness enables *resilience*; this is its goal and raison d'être. The logic of resilience is peculiar in that it aims not precisely at a future that is "better" in any absolute sense but rather at a smart infrastructure that can absorb constant shocks while maintaining functionality and organization. Following the work of Bruce Braun and Stephanie Wakefield, we describe resilience as a state of permanent management that does away with guiding ideals of progress, change, or improvement.[27]

The term "resilience" plays important, though differing, roles in multiple fields. These include engineering and material sciences: since the nineteenth century, the modulus of resilience has measured the capacity of materials such as woods and metals to return to their original shape after impact. Resilience is also an important term in ecology, psychology, sociology, geography, business, and public policy, in which it names ways in which ecosystems, individuals, communities, corporations, and states, respectively, respond to stress, adversity, and rapid change.[28] However, the understanding of resilience most crucial to smartness and the smartness doctrine was first forged in ecology in the 1970s, especially in the work of C. S. Holling, who established a key distinction between stability and resilience. Working from a systems perspective and intrigued by the question of how humans could best manage elements of ecosystems that were of commercial interest (e.g., salmon, wood, etc.), Holling developed the concept of resilience to contest the premise that ecosystems were healthiest when they returned quickly to an equilibrium state after being disturbed (and in this sense his paper critiqued then-current industry practices).

Holling defined *stability* as the ability of a system that had been perturbed to return to a state of equilibrium, but he argued that stable systems were often unable to compensate for significant, swift environmental changes. As Holling put it, the "stability view [of ecosystem management] emphasizes the equilibrium, the maintenance of a predictable world, and the harvesting of nature's excess production with as little fluctuation as possible." However, he continued, this approach cannot take into account that "a stable maximum sustained yield of a renewable resource might so change [the conditions of that system] . . . that a chance and rare event that previously could be absorbed can trigger a sudden dramatic change and loss of structural integrity of the system."[29] Resilience, by contrast, denoted for Holling the capacity of a system *to change* during periods of intense external perturbation and thus a capacity to persist over much longer time periods than in the case of stable systems. The concept of resilience encourages a management approach to ecosystems that "emphasize[s] the need to keep options open, the need to view events in a regional rather than a local context, and the need to emphasize heterogeneity" (see figure I.5). Resilience is in this sense linked to concepts of crisis and states of exception, for resilience is a virtue only when the latter are assumed to be quasi-constant. Holling also underscored that the movement from stability to resilience depended upon an epistemological shift: "Flowing from this [understanding of resilience] would be not the presumption of sufficient knowledge, but the recognition of our ignorance: not the assumption that future events are expected, but that they will be unexpected."[30]

Smartness abstracts the concept of resilience from ecology and turns it into an all-purpose epistemology and value, positing resilience as a more general strategy for managing perpetual uncertainty in all fields and encouraging the premise that the world is indeed so complex that unexpected events are the norm. Smartness enables this generalization of resilience in part because it abstracts the concept of populations from the specifically biological sense employed by Holling: in addition to populations of individual organisms, smartness also sees populations of preferences, traits, and algorithmic solutions. Resilience also functions in

I.5 An example of resilient design: experimental floating wetlands on the Charles River, Boston, designed to suppress algal blooms. Constructed by Northeastern University. *Source:* Photo by Orit Halpern, December 30, 2020.

the discourse of smartness to collapse the distinction between *emergence* (something new) and *emergency* (something new that threatens). By collapsing this distinction, resilience produces a world in which any change purportedly can be technically managed and assimilated by maintaining the ongoing survival of the system rather than the survival of individuals, or even particular groups of individuals. Smartness thus focuses on the management of the *relationships between* different populations of data, some of which can be culled and sacrificed for systemic maintenance.[31] In doing so, resilience is a key functionary in what Jennifer Gabrys has called "the becoming environmental of computing" and in what Benjamin Bratton has labeled "planetary scale computing."[32] Smartness makes the environment into a medium while explicitly transforming evolution. Planned obsolescence and preemptive destruction combine here to encourage the introduction of ever more computation into the environment, as well as emphasize that resilience of the species may necessitate sacrifices of "suboptimal" populations.

The discourse of resilience effectively erases the differences among past, present, and future. Time is understood not through a historical or progressive schema but rather through schemas of repetition and recursion (the same shocks, and the same methods, are repeated again and again), even as these repetitions and recursions produce constantly differing territories. This is a self-referential difference only measured or understood in relation to the many other versions of smartness (e.g., earlier smart cities), which all tend to be built by the same corporate and national assemblages.

The collapse of emergence into emergency also links resilience to financialization through derivation, as the highly leveraged complex of Songdo already demonstrated.[33] The links that resilience establishes among emergency, financialization, and derivatives is also exemplified by New York City, which, after the devastation of Hurricane Sandy in 2012, adopted the slogan "Fix and Fortify." This slogan underscores an acceptance of future shock as a necessary reality of urban existence while at the same time leaving the precise nature of these shocks unspecified (though they are often implied to include terrorism as well as environmental devastation). The naturalization of this state is vividly demonstrated by the irony that the real destruction of New York had earlier been imagined as an opportunity for innovation, design thinking, and real-estate speculation. In 2010, shortly before the real hurricane hit New York, the Museum of Modern Art and PS1 ran a design competition and exhibition titled *Rising Currents*, which challenged the city's premier architecture and urban design firms to design for a city ravaged by rising sea levels as a result of global warming:

MoMA and PS1 Contemporary Art Center joined forces to address one of the most urgent challenges facing the nation's largest city: sea-level rise resulting from global climate change. Though the national debate on infrastructure is currently focused on "shovel-ready" projects that will *stimulate the economy*, we now have an important *opportunity* to foster *new research and fresh thinking* about the use of New York City's harbor and coastline. As in past economic recessions, construction has slowed dramatically in New York, and much of the city's *remarkable pool of architectural talent is available* to focus on innovation.[34]

It is difficult to imagine a clearer statement about the ideal relationship of urban planners to crisis: planning must simply assume and assimilate future, unknowable shocks, and these shocks may come in any form. This rather stunning statement turns economic tragedy, the unemployment of most architects, and the imagined coming environmental apocalypse into an opportunity for speculation (with *speculation* understood to be simultaneously a technical, aesthetic, and economic operation). This is a quite literal transformation of emergency into emergence and of creating a model for managing perceived and real risks to the population and infrastructure of the territory not by "solving" the problem but by absorbing shocks and modulating the ways in which the environment is managed. New York in the present becomes a mere demo for postcatastrophe New York, and the differential between these two New Yorks is the site of financial, engineering, and architectural interest and speculation.

This relationship of resilience to the logic of demos and derivatives is illuminated by the distinction between risk and uncertainty first proposed in the 1920s by the economist Frank Knight. According to Knight, uncertainty, unlike risk, has no clearly defined end points or values.[35] It offers no clear-cut terminal events. If the geopolitical dynamics of the Cold War understood nuclear testing and simulation as a means of avoiding an unthinkable but nonetheless predictable event—nuclear war—the formula has changed; we now live in a world of fundamental uncertainty, which can only ever be partially and provisionally captured through discrete risks. When uncertainty, rather than risk, is understood as the fundamental context, "tests" can no longer be understood primarily as a simulation of life; rather, the test bed makes human life itself an experiment for uncertain technological futures. Uncertainty thus embeds itself in our technologies, both of architecture and finance. In financial markets, for example, risks that are never fully accounted for are continually "swapped," "derived," and "leveraged," in the hope that circulation will defer any need to actually represent risk, and in infrastructure, engineering, and computing, we do the same.[36]

As future risk is transformed into uncertainty, smart and ubiquitous computing infrastructures become the language and practice by which to imagine and to create our future. Instead of looking for utopian answers to our questions regarding the future, we focus on quantitative and algorithmic methods and on logistics; on how to move things from point A to point B rather than questions of where they *should* end up (or whether they should be there at all). Resilience as the goal of smart infrastructures of ubiquitous computing and logistics becomes the dominant method for engaging with possible urban collapse and crisis (as well as the collapse of other kinds of infrastructure, such as those of transport, energy, and finance). Smartness thus becomes the organizing concept for an emerging form of technical rationality, the primary goal of which is management of an uncertain future through a constant deferral of future results; for perpetual evaluation through a continuous mode of self-referential data collection; and for the construction of forms of financial instrumentation and accounting that no longer engage, or even need to engage, with what capital extracts from history, geology, or life.

GENEALOGIES

Each of the four chapters in this book focuses on one of the following terms—"populations," "experimental zones," "optimization/derivation," and "resilience"—and provides a genealogy of the concepts, techniques, and technologies that led to the present function of these concepts and their associated technologies within the smartness mandate. As will be evident in our chapters, each term emerged, and was engaged, within multiple discourses and technologies, including ecosystem ecology, evolutionary biology, management science, computer science, and economics, to name just a few. There was, in addition, often significant cross talk and conceptual and technical borrowing among these disciplines. On the one hand, this complexity makes a complete, or comprehensive, genealogy of smartness difficult and perhaps even impossible. On the other hand, this complexity underscores the need for a mapping of the sort provided by this book. However, we do not consider our account to be the only possible genealogy of smartness, and we can imagine other genealogies that focus on different authors, engineers, and techniques. Though our genealogy is intended to illuminate the deep logic of smartness—a logic that would also apply to alternate genealogical accounts—we employ excurses in each chapter to gesture both toward the fact that our account is one of several possible ways to explain the rise of the smartness mandate and that the current smartness mandate was *not* the only possible outcome of the techniques and concepts we describe.

In constructing our genealogy of the smartness mandate, we drew on earlier work in the history of science, science and technology studies (STS), media studies, and urban/design studies. We found especially helpful the work of historians of science and STS scholars who have focused on the history of cybernetics and on the histories of changing scientific conceptions of rationality.[37] We drew inspiration from Paul N. Edwards's *A Vast Machine: Computer Models, Climate Data, and the Politics of Global Warming*, which documents the multiple scientific techniques, discourses, and political projects that became linked to enable *climate modeling*.[38] Equally important to us have been the histories of environmentality and environment developed by scholars such as Peder Anker (and especially the links he draws among ecology, architecture, cybernetics, and empire), as well as Etienne Benson's work on *environmentalisms*, which underscores both the historical contingency of definitions of environment (and therefore also of models of environment and the types of actions understood as typical of environments) and the relationship of these definitions to media.[39]

Indeed, a core theme in this book is the transformation of *environment* into a media surround that is also a political ecology, to borrow from Fred Turner.[40] Just as climate is both the product and producer of media, the control of climate is also about the control of populations, as Yuriko Furuhata, Nicole Starosielski, and Daniel Barber have shown. How climate is managed, whether by means of air-conditioning or by building management systems that are smart, depends upon premises about the human subject, about the norms of the body, and about how social order can be organized through spatial relations. In these cases, climate as medium is also climate as biopolitics and relations of power. These relations of power include both colonial and postcolonial relations and relations between the Global North and South. As these authors have argued, evolving understandings of climate, environment, population, and media were central to the postcolonial and post–World War II global organization of power and territory.[41] In their work on drone warfare and global media infrastructures, Lisa Parks and Caren Kaplan have further argued that these media infrastructures are *"biopolitical machines* that have the potential to alter life in a most material way." These machines are historically and culturally situated and emerge from histories of militarization and conquest that influence not only the forms of politics but also the strategies that emerge from these machines.[42] In similar fashion we understand smartness as emerging at a particular time and from particular histories, especially that of neoliberalism. We hope in this book to document, at least in limited ways, the many valences of smartness and the place of histories of empire and coloniality in structuring contemporary regimes of digital smartness.

As will be evident in this book, smartness is both an idea and an infrastructure. As Shannon Mattern argues, we must attend to "the hardware of media"; that is, we need to attend to the many materialities and histories of media infrastructure. This means understanding not only that digital media is specific and has its own forms but also that the overwhelming focus on questions of signals or communication in media studies sometimes comes at the cost of engaging different subjects and materialities. Similarly, to focus only on visible infrastructures, such as roads, or sewers, or fiber-optic cables, can come at the cost of recognizing the force of concepts, ideas, and imaginaries that enable and flow from them. Mattern develops the idea of media archaeology as literal engagement of digging up pasts as one resolution of this problem.[43] We develop a similar approach here by taking seriously the point that communications media have histories and shape territorial forms while at the same time attending just as seriously to issues of materiality (mining, extraction, algorithms) and the ideas that often predate, and encourage, the construction of smartness and the penetration of smartness into the environment.

Since we engage so many disciplines and trace these across a fairly lengthy time frame (roughly from the 1930s to the present), it will be helpful to note here that each of our chapters outlines a similar temporal rhythm. In each chapter a set of technical and theoretical tools first developed in the cybernetic sciences around and immediately after World War II was then reframed drastically in the 1970s (a period of global political turmoil but also increasing computational capacities). This reframing was then worked out more fully in the 1980s and 1990s and took on its contemporary form in the early 2000s, as computational speed and spread enabled what had once been only a dream—namely, environmental sensing and computing—to become a reality.

Chapter 1, "Smartness and Populations," begins with the theorization in the 1930s of what geneticist Ernst Mayr called *population logic* and a parallel emphasis on the importance of individual differences in economist Friedrich Hayek's theory of markets; the key innovation for both was to understand populations or markets as entities that learned, at least in a sense. These understandings of populations as sites of learning were brought into computational models of learning in the late 1950s and 1960s. This approach to learning was further cemented but given a decidedly market-oriented twist in the 1970s by distinguishing itself from a competing theory of the link between populations and computing instantiated in the famous report *The Limits to Growth*.[44] This report relied on the computer modeling of world systems but presented both markets and populations as dumb (i.e., incapable of learning). The market-oriented approach to learning populations was integrated in the early 1990s into a series of internet applications, such as the Google PageRank algorithm, and has since become a widespread principle of linking individuals by means of sensing and computing.

Chapter 2, "Demo or Die: The Zones of Smartness," takes up the territory of smartness (experimental zones) and employs the theory of "soft architecture" that Nicholas Negroponte developed in the 1970s as a key lens for understanding the link between experimentation and territory that is central to the zonal logic of the smartness mandate. We emphasize that Negroponte, who was based at the Massachusetts Institute of Technology (MIT), relied on approaches to computing first developed in post–World War II cybernetics discourse, which included Oliver Selfridge's "Pandemonium" model of computer learning, Jay Forrester's systems approach to urban dynamics, and other MIT-linked attempts to model urban change. Negroponte's innovation was to apply these learning approaches to the *design* of urban infrastructure, with the goal of optimizing the learning capacities of the populations within cities. This approach subtly reframed the zoning principles upon which cities had been planned since the early twentieth century by focusing on transforming urban centers into sites of perpetual demos, or experimental zones.

Chapter 3, "Derivation, Optimization, and Smartness," explores the key means by which smartness produces learning from distributed populations—namely, by deriving value from what was earlier understood to be noise and waste. As in the case of chapter 2, we begin in the midpoint of our genealogy, the 1970s, focusing on the development of a new financial tool (the Black-Scholes option pricing equation) and underscoring the importance of noise and systemic connections for this technology. We note that although the Black-Scholes option pricing equation may seem esoteric and limited to finance, in fact it exemplifies a basic logic that is operative in a wide variety of contemporary technologies, including "platforms," such as Uber and Airbnb; cognitive "mining" technologies; and population-level biobanks. We then trace the origins of this approach to noise and waste back to the post–World War II period, focusing especially on the psychologist Donald Hebb and the management theorist Herbert Simon.

Chapter 4, "Resilience," focuses on the goal of smartness—namely, to enable resilience. Here, too, our genealogy begins in the period around World War II as the new discipline of ecosystem ecology developed a set of tools for understanding how natural environments respond to shocks from their outsides, such as radioactive fallout from nuclear bombs. Yet where ecosystem ecology still prioritized stability and homeostasis, ecologist Holling developed his concept of resilience in the 1970s as a way to center instability and perpetual change as the basic rule for ecosystems. Holling's theory of resilience was intended to provide a model for managing ecosystems but quickly became a more general model of management itself. By the early 2000s, this model implied that management was first and foremost a matter of developing flexible systems that, through data-intensive but selective surveys of their environments, could quickly adjust to whatever new shock the environment might throw at them.

In our coda we contemplate the ways that ecology, economy, and technology have been reorganized through the mandate to make our world smart. In this final moment, we return to the Chilean Atacama to ruminate on how new forms of population-grounded perception and cognition might offer opportunities to make new worlds that are more just, equitable, and imaginative than those currently constrained through the limited comprehension of smartness often propagated by large-scale developers and technology industries.

SMARTNESS AND CRITIQUE

As we hope is clear from our account of the smartness mandate above, smartness is both a reality and an imaginary, and it is this commingling that underwrites both its logic and the magic of its popularity. Consequently, a critique of smartness cannot be simply a matter of revealing the inequities produced by its current instantiations. Critique is itself already central to smartness, in the sense that perpetual optimization requires perpetual dissatisfaction with the present and the premise that things can always be better. Therefore, the advocates of smartness can always plausibly claim (and likely also believe) that the *next* demo will be more inclusive, equitable, and just. A critique of smartness thus needs to confront directly the terrible, but necessary, complexity of thinking and acting within earthly scale—and even extraplanetary scale—technical systems.

On the one hand, this means stressing the ways in which the smartness mandate blunts what might otherwise be understood as the urgency of conditions of environmental degradation, inequality and injustice, mass extinctions, wars, and other forms of violence via the demand that we understand our present as a demo oriented toward the future and (as a consequence) by encouraging us to employ a single form of response—namely, increased penetration of computation into the environment—for all crises. On the other hand, it is impossible to deny not only the agency and transformative capacities of smart technical systems but also the deep appeal of this approach to managing an extraordinarily complex and ecologically fragile world. (And none of us is eager to abandon our cell phones or computers!) Moreover, the epistemology of partial truths, incomplete perspectives, and uncertainty with which Holling sought to critique capitalist understandings of environments and ecologies still holds a weak messianic potential for revising older modern forms of knowledge and for building new forms of affiliation, agency, and politics grounded in uncertainty, rather than objectivity and certainty, keeping us open to plural forms of life and thought. However, insofar as smartness separates critique from conscious, collective human reflection—that is, insofar as smartness seeks to steer communities algorithmically, in registers operating below consciousness and human discourse—critiquing smartness is in part a matter of excavating and rethinking each of its central concepts and practices (experimental zones, populations, optimization, and resilience) and the temporal logic that emerges from the particular way in which smartness combines these concepts and practices.

1

SMARTNESS AND POPULATIONS

Many commentators have drawn attention to the parallels between the COVID-19 viral pandemic that began in 2020 and the Spanish flu pandemic that occurred almost precisely a century earlier in 1918–1920. However, the differences are arguably much more striking, not least because COVID-19 spread and was countered by means of governance strategies developed within a smart world. In the case of the early twentieth-century Spanish flu, many European governments hid the extent of the illness during the early months of the pandemic, leading to the erroneous description of the flu as Spanish in origin (Spain was simply the first country that did not censor information about the flu). In the case of COVID-19, by contrast, information about the disease was—or at least was imagined to be—constantly available via a 24-hour, internet-mediated global news network, and the specific agent of infection was identified and its genetics sequenced within a week of its emergence in Wuhan, China. While the initial lockdown of Wuhan depended on centuries-old techniques of physically controlling roads, opening up the city a few months later relied on a smartphone app that indicated, via a green symbol, which citizens were free to resume normal activity within the city. Companies such as Google and Apple developed their own versions of such apps, intended as optional, rather than required, of citizens in Western European countries and the United States. Many Western European and North American countries sought to justify social-distancing measures to their citizens by publishing interactive epidemiological "curve-flattening" computer simulations and allowed residents to track in real time information such as the number of infections, recoveries, deaths, hospital beds occupied, and hospital capacities. Schools and universities replaced in-person classes with online learning systems that employed algorithms to determine the actual and target skills for millions of new users, and companies such as Netflix and Amazon were able to take advantage of consumers suddenly forced to stay at home for months both to increase their business revenues and to refine their consumer prediction engines.

Drawing on her earlier concept of the shock doctrine, Naomi Klein offered a perceptive account of the ways in which companies committed to smart technologies took advantage of the COVID-19 pandemic to advance their interests. She noted, for example, New York governor Andrew Cuomo's appointment of former Google CEO Eric Schmidt to a commission to develop strategies for New York's recovery from the economic devastation produced by COVID-19 and Cuomo's related partnership with the Bill and Melinda Gates Foundation to "develop 'a smarter education system,'" which of course meant one more mediated by smart technologies. Klein stressed that though this smart future was presented as a solution to the specific problems engendered by COVID-19, in fact this vision of a "precise app-driven, gig-fueled future" had already been promoted by these same individuals years prior to the pandemic. Klein contended that public opinion had begun to turn against that vision, exemplified by the fact that "presidential candidates were openly discussing breaking up big tech," "Amazon was forced to pull its plans for a New York headquarters because of fierce local opposition," and "Google's own workers were refusing to build surveillance tech with military applications." However, Klein suggested, the pandemic proved to be a crisis sufficiently disorienting for the general public that these doubts about smart technologies could be swept away. Klein described this operation as "the Pandemic Shock Doctrine," by which she meant that, in the same way that neoliberal economists in the 1970s and 1980s made use of the psychological shock and distraction engendered by (often deliberately constructed) social "crises" to advance neoliberal economic reforms, advocates of smartness were now relying on the shock and distraction of a world-wide pandemic to create what she called the "Screen New Deal."[1]

While Klein astutely described the ways in which advocates of smart-ness employed the COVID-19 pandemic to advance their interests, her effort to understand this expansion and intensification of smart technolo-gies through the concept of the shock doctrine obscures more than it illu-minates. As Klein documented in her book of the same name, the shock doctrine denoted the linkage of a specific physiopsychological theory of shock with neoliberal theories of the market. Klein located the remote origin of this theory of shock in neurologist Donald O. Hebb's account, in the 1940s, of the ways in which neurons become linked to one another in the context of learning experiences, but Klein stressed that Donald Ewen Cameron—who was employed by the CIA and was interested in the ways in which Hebb's theory of neuronal learning implied the possibility of "rewiring" people through techniques such as torture—was the real source of the shock doctrine. As we note in this chapter and in chapter 3, there is an important historical connection between Hebb's approach to learning and the smartness mandate. Yet the connection runs through concepts of *populations* rather than concepts of individualized shock. Smartness is a practice of governance oriented toward populations, whether of human beings, neurons, or computer models (to name a few of the possibilities), and to miss this dimension of the Screen New Deal is to misunderstand fundamentally the premises and operations of smartness. Whereas the concept of shock presumes a displacement from a widely shared norm that otherwise persists for long periods of time, smartness, by contrast, values divergences from such a norm and presumes constant change and volatility.[2]

In this chapter we document the ways in which the concept, technolo-gies, and governance practices of smartness emerged in part through the hybridization of models and governance mechanisms of three different approaches to populations. The first part of this chapter distinguishes analytically among these three approaches. The first approach, which we exemplify through the work of Thomas Malthus, premises its governance strategies on the principle that populations are made up of essentially *homogeneous* individuals who are driven by unalterable natural drives;

from this perspective, a population is incapable of anything like learning, and population governance means developing techniques that can regulate the otherwise uncontrolled dynamics of growth and contraction. The second strategy, which we exemplify with late nineteenth-century actuarial insurance practices, focuses not on managing the overall growth or expansion of a population but rather on locating and exploiting differences within a population—for example, age brackets or different occupations—to manage collective risk. The third strategy, which we exemplify both by means of biologist Ernst Mayr's mid-twentieth-century concept of *population thinking* and by economist Friedrich Hayek's theory of markets, understands populations (or markets, in Hayek's case) as entities that *learn* (or at least evolve) by responding to changes in the environment.

Although the Malthusian approach to populations was developed earlier than the actuarial approach, which was itself developed earlier than the understanding of populations as agents of learning, the first part of this chapter does not tell a story of the displacement of one population approach by the next, and we emphasize this point by briefly stressing ways in which the first two approaches have remained central to contemporary governance practices. Rather, the first part of the chapter illuminates elements—namely, these population models and their associated techniques of governance—that were brought together in the formation of the smartness mandate. The second part of our chapter shows how these elements, which remained largely disconnected from one another until the 1970s, began to interact, with computation serving as the medium of their connection. We note that although Mayr's and Hayek's population approaches had quickly migrated into computer science in the 1950s and 1960s, it was the hugely influential report *The Limits to Growth* (1972) that brought together the Malthusian model with world computer-modeling techniques and served as a key impetus for the further hybridization of these models and techniques.

THREE MODELS OF POPULATION

(DUMB) MALTHUSIAN POPULATIONS
Since the seventeenth century, concepts of population have been an important element of Western governance practices, and Francis Bacon's

seventeenth-century *new science* established many of the basic coordinates for the subsequent importance of this term. Bacon urged seventeenth-century legislators to see the state of "the Population" as a key factor in encouraging or hindering political upheaval. However, Bacon also stressed that a population is not "to be reckoned only by number; for a smaller number that spend more, and earn less, do wear out an Estate sooner than a greater number that live lower, and gather more."[3] Bacon thus marked out two fundamentally different models for understanding and managing populations. On the one hand, populations could be understood as a large mass of homogeneous individuals, with the key variable as the positive or negative rate of growth of that mass. On the other hand, Bacon stressed the importance of differences among the members of a population. In Bacon's example these were differences in what we now call the economic sphere ("spending, earning, and gathering"), but they could also be differences in other areas of life, such as occupation, age, or illness.

For most of the eighteenth century, European legislators focused primarily on the first model—a population understood as a large collection of essentially homogeneous individuals—and tended to assume that the larger the population of an area, such as England, the greater its political power. However, Thomas Malthus's *A Principle of Population* (1798), while drawing on this same approach, fundamentally altered this concept of population in ways that continue to structure our present. Malthus's key innovation was to view the growth of population not as an intrinsically good attribute of the state but rather as a *threat* that needed to be managed. Malthus supported this approach by arguing that populations, whether of plants, animals, or humans, "when unchecked, increased in a geometrical [i.e., exponential] ratio."[4] For Malthus, populations blindly sought to cover as much geographic space as possible, and he illustrated the idea of unregulated population growth by means of an image drawn from Benjamin Franklin's account of the spread of plants:

Were the face of the earth, [Franklin] says, vacant of other plants, it might be gradually sowed and overspread with one kind only; as, for instance, with fennel: and were it empty of other inhabitants, it might in a few ages be replenished from one nation only; as, for instance, with Englishmen.[5]

That the globe has not been overrun by fennel was a consequence of the fact that fennel was a source of food for other populations and competed with other plants for space. As a consequence, Malthus contended, the food sources for animal populations can only increase "in arithmetical ratio." As a population starts to overshoot its available food sources, population growth is then always checked by forces such as starvation or illness.

Malthus was a political economist, and his theory of population was intended to intervene in contemporary debates about government poor relief. He argued that providing food or money to the poor was ineffectual, as this simply delayed the point at which the "excess" population would end up starving. For Malthus, the drive for reproduction that characterized populations could not be altered, and it was pointless to try to change this fundamental behavior of a population. That is, no population—whether of plants, animals, or humans—had any capacity for learning, and the best one could do was adopt legislative measures that discouraged a human population's blind drive toward overexpansion.[6]

Malthus's understanding of population as an ever-present threat that had to be regulated by economic and political means was tremendously influential in his own lifetime and has had a long (and still continuing) afterlife in multiple fields. Malthus's theory remained fundamental, for example, for nineteenth-century political economists and social theorists.[7] In the early twentieth century, mathematicians and physicists such as Alfred James Lotka and Vito Volterra developed more mathematically sophisticated "predator-prey" models for the population dynamics originally proposed by Malthus and in this way helped move the Malthusian paradigm into sciences such as ecology.[8] Although, as we will note below, Malthus's theory led Charles Darwin to a quite different theory of population, *social Darwinists* often mapped Malthusian worries about competition between populations onto social divisions (for example, purported racial differences among groups of humans).[9] We will also describe below the extent to which Malthus's approach to population was central to the *population debate* that began in the 1960s, was connected to global computer modeling in *The Limits to Growth* (1972), and was instantiated in both the Chinese one-child policy and US Agency for International Development (USAID) requirements to limit world growth. Even as the Chinese one-child policy was phased out in the early twenty-first century, the Malthusian approach to world population has remained central to discussions of how to address the effects of global warming.

ACTUARIAL POPULATIONS

Where the Malthusian approach treats populations primarily as a single number, a different approach to populations developed in the eighteenth and nineteenth centuries within discourses focused on the economic management of risk, especially insurance.[10] While there is a long Western European history of insurance—including maritime insurance, which insured commercial investors against losses of sailing ships or cargo, and fire insurance, which insured against the loss of property in increasingly dense urban environments—up until the eighteenth century the basis for determining premiums depended primarily on rules of thumb and the judgment of the insurer.[11] However, in the eighteenth century increasing numbers of groups formed to provide life insurance for individuals, and in these cases premiums were often based on relatively large data sets about deaths in a local population. As many of these early insurers discovered (often the hard way), both accurate and large population data sets—as well as careful use of what would eventually become known as statistical mathematical techniques—were necessary to set individual premiums high enough so that the insurance company did not end up paying out more in claims than it received in premiums.

In the late nineteenth and early twentieth centuries, some insurance functions, such as workplace accident insurance, were either taken over by the state (e.g., in Germany) or were legally required of employers (in the US). The establishment of collective "security" through the use of population-level statistics became a central principle of the modern welfare state, as well as its US variant, which took the form of institutions developed during the New Deal period, such as Social Security.[12]

The actuarial approach to populations is based on three key premises. First, and in contrast to the Malthusian emphasis on population expansion or contraction, the actuarial approach deals best with a steady-state population that can be divided into multiple groups by means of categories such as age, sex, and occupation. This approach requires much more extensive methods and categories of data collection than the Malthusian approach, which focuses more or less exclusively on births and deaths. Second, the actuarial approach presumes that although it is impossible to follow or predict each individual's unique path through every relevant category (illnesses, injuries, occupations, etc.), statistical knowledge of the entire population allows one to ignore these individual differences in favor of larger-scale regularities. For example, given knowledge of the specific accidents that have occurred to all the individuals within a sufficiently large population, one can then generate relatively certain knowledge about the frequency of specific *kinds* of accidents within this population. Third, this statistical knowledge of the frequency of different classes of, for example, workplace accidents then enables an insurer to set a premium for individuals (i.e., to quantify risk monetarily). Quantifying risk in this way means that it no longer matters whether a *specific* insured individual has a specific kind of accident; all that matters is that the statistical regularities observed in the history of this population continue to hold in the future.

Just as the Malthusian approach to population has remained a vital part of statecraft and governance structures into the twenty-first century, the actuarial approach to populations has ramified into multiple discourses. It continues to serve as the foundation of both state-sponsored and private health insurance (and was, as we will note in chapter 3, central to the US housing crisis of 2007, which spread into a world economic crisis). More generally, the actuarial approach tends to emerge whenever *risk* categories are connected to questions of populations, whether or not insurance per se is at issue.

The relationship between risk and an actuarial approach to populations, for example, is evident in the early twentieth-century emergence of the project of *personalized medicine*—also sometimes described as smart medicine—though this project has no direct connection to insurance. Since the mapping of the human genome in 2000, some medical researchers have seen genetic information as a means for creating different actuarial-like categories for disease risk. Rather than assuming one norm of human physiology and behavior, researchers are increasingly interested in establishing differing genetic risk factors for individuals. Establishing risk factors requires statistical correlations among genetic profiles, lifestyle choices, and health events, which then enable predictions such as the following: "Those with gene variant X who smoke and who exercise fewer than 20 hours a week have a 20 percent higher chance of developing condition A than someone who lacks the mutation of gene X, does not smoke, and exercises 20 or more hours a week." Risk factors are intrinsically relativistic:

individual A's risk of developing condition X is not determined against a single norm but in relationship to other genetic profiles, lifestyle choices, and health events. In order to determine the risk factors noted above, researchers need access to huge numbers—on the order of tens of thousands or, if possible, millions—of samples of blood or tissue that can be linked to health records and data about living environment and lifestyle choices.

To meet this demand for tissue and information, many private and public biobanks have emerged in the last decade. Biobanks have become more valuable as data associated with a specimen can be continually updated with new, vital facts, such as the emergence of a new sickness or a new lifestyle choice (e.g., starting or ending smoking). As a consequence, some university research hospitals are developing protocols for linking clinical health-care records to biobanks so that each time an individual visits the clinic, biobank data is updated with that new information.[13] The hospital clinic is in this sense repurposed into a medium capable of capturing, both materially and informatically, individual variations in large numbers of individuals, which in turn enables a feedback loop between medical research institutions and the health of the larger population. By serving as a key node of the biobank, the clinic not only remains a site at which the results of medical research are applied in the form of diagnoses, therapies, and medicines but also becomes a site through which medical research is generated.[14]

Although the practice of linking risk, clinical information, and environmental factors dates back to population studies conducted in the 1950s, biobanks mark a new threshold in the use of actuarial population logic, both because of the size of the biocollections and the emphasis of biobanks on something "internal" (genetic information) that cannot be determined outside of the clinic.[15] Rather than restricting themselves to small, "representative" samples of a population, biobanks seek to include hundreds of thousands of participants; that is, they seek participation at the level of entire populations. The fact that biobanks base health risk factors on genetic or other internal forms of information means that individuals cannot determine risk factors without the diagnostic tools of the genetics lab. Knowing that one is, for instance, female, Black, and a smoker and lives in a stressful inner-city environment is no longer enough to determine one's risk factor for diabetes; one's DNA sample must also be sent to a lab so that one's genetic profile can be integrated into that risk factor equation.

The emphasis of personalized medicine on *lowering* risk also tends to encourage a far less nuanced—and ultimately paradoxical—desire to approximate an ideal type: namely, an individual with the lowest-possible risk for all known conditions. Personalized medicine reimagines patients as financial investors: each individual inherits a portfolio of biological "investments," and the patient/investor should seek to minimize exposure to risk. Although some risks can be minimized by lifestyle choices, such as diet and exercise, personalized medicine is more oriented toward managing risks through pharmaceuticals. However, since there is no longer a single "norm" toward which an individual could orient their risk profile, there is in principle no end to the number of drugs an individual might take to lower exposure to risk even further.[16]

POPULATIONS AS AGENTS OF LEARNING: MAYR AND HAYEK

For both the Malthusian and actuarial approaches, a scientific observer can generate knowledge *about* a population—for example, the rate of growth for a population or the statistical likelihood of specific events within a population—but the population itself has no internal capacity for responding to its environment and, hence, learning from experience. However, in the 1940s and 1950s, a new model of populations—namely, as entities that *were* capable of problem-solving and therefore learning, at least of a sort—emerged in both evolutionary biology and in economics. In the two instances we consider here, evolutionary biologist Ernst Mayr's approach to speciation and Friedrich Hayek's theory of markets as information processors, differences between individuals were the key to a population's learning abilities.

Ernst Mayr, an extraordinarily influential twentieth-century biologist, was central in bringing about the so-called modern synthesis of Mendelian genetics, taxonomy, and the Darwinian theory of evolution that still serves as the basis for essentially all genetically oriented approaches to living beings. As we noted above, Charles Darwin drew explicitly on Malthus's model of population, contending in *On the Origin of Species* (1859) that readers should understand his book as "the doctrine of Malthus, applied to the whole animal and vegetable kingdoms."[17] More precisely, Darwin drew from Malthus the image of individuals competing with one another for resources. However, Darwin was interested in the reasons that one individual would be more successful than another in this fight for scarce resources, and he attributed this to differences, or *variations*, among individuals. Darwin transformed Malthus's image of population as a collection of essentially identical individuals into an image of population as a collection of individuals who each differ from the other members of the population.

In the early to mid-twentieth century, researchers following in the Darwinian tradition focused increasingly on genetics as the primary source of these individual differences, and Mayr was a key figure in this development. Mayr was by training an ornithologist, specializing in the birds of Oceania and Indonesia, and his contribution to the modern synthesis focused especially on aligning a theory of speciation (the processes by which new species emerge) with methods of genetic research developed in the early and mid-twentieth century. Mayr argued that species should be understood as geographically distributed collections of populations, each of which has the possibility of becoming a new species, given the right conditions.

Mayr's emphasis on populations was in part a function of his scientific specialty. As a taxonomist of bird species, Mayr's work involved classifying both living and dead instances of birds into species and subspecies. In both the field and in his position at the American Museum of Natural History, this meant seeking to classify individual instances of different bird species although the physical features, such as body length and plumage color, might vary considerably among individuals as a function of age, sex, season, and geographic location. However, as Mayr stressed in his first major book, *Systematics and the Origin of Species: From the Viewpoint of a Zoologist* (1942), "birds are better known taxonomically than any other class of animals," for ornithologists had access to an enormous number of samples.[18] Mayr noted, for example, that in 1941, the bird collection in the American Museum of Natural History comprised "about 800,000 skins, or 100 skins per species, and about 30 specimens per subspecies"; in addition, taxonomists could request loans of specimens from other institutions.[19] Hence, it was not uncommon for a taxonomist to examine many thousands of specimens of a species, which meant hundreds of examples of individual subspecies, which Mayr also later described as populations.[20] Ornithologist taxonomists such as Mayr were thus, by the very nature of their job, oriented simultaneously toward large populations of samples and the visible differences between each sample. The new science of twentieth-century genetics underscored that these differences among individuals at the level of the phenotype also held true at the level of genetics, for each individual was a unique combination of genes.

For Mayr, speciation should be understood as a process in which genetic differences among members of a local population exposed to new environmental conditions cause that population to diverge from the original species of which it had been a part. This process was especially clear in cases in which one population of a species became geographically isolated—for example, on an island—from other populations of the species. The geographically isolated population encountered a different environment than other populations of the same species. As a consequence, individuals who might have been less fit in the original habitat could emerge as more fit in the new environment, enabling a significant shift in the "normal" characteristics of the island population. Over the long term, the island-locked population could develop into a new species that could no longer breed with other populations of the original species (see figure 1.1). The concept of population was thus for Mayr a means for focusing attention on the necessary conditions for *innovation* or *novelty* (in this case, the way in which a new species emerges from an older species).

Mayr's interest in populations as sources of biological innovation encouraged him to stress repeatedly a point that he felt was often not taken seriously enough by evolutionary biologists: namely, that every member of a population was biologically unique. Mayr underscored this point by distinguishing between what he called "typological" thinking and "population" thinking. The "assumptions of population thinking," Mayr wrote,

are diametrically opposed to those of the typologist. The populationist stresses the uniqueness of everything in the organic world. What is true for the human species,—that no two individuals are alike,—is equally true for all other species of animals and plants. . . . All organisms and organic phenomena are composed

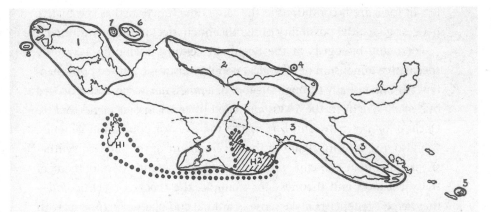

FIG. 15. The insular races (4-8) of the New Guinea kingfisher *Tanysiptera galatea* have developed almost specific rank in their small isolated ranges. The three mainland races (1-3) are very similar to each other. 1 = *galatea;* 2 = *meyeri;* 3 = *minor;* 4 = *vulcani;* 5 = *rosseliana;* 6 = *riedelii;* 7 = *carolinae;* 8 = *ellioti.* Range expansion of *minor* into south New Guinea has led to an overlap with *hydrocharis* (H₁ and H₂) which was formerly isolated by an arm of the sea.

1.1 Ernst Mayr's illustration of speciation by means of geographic isolation. *Source:* Ernst Mayr, *Systematics and the Origin of Species from the Viewpoint of a Zoologist* (Cambridge, MA: Harvard University Press), 153, fig. 15. Copyright © 1942, 1970 by Ernst Mayr.

of unique features and can be described collectively only in statistical terms. Individuals, or any kind of organic entities, form populations of which we can determine the arithmetic mean and the statistics of variation. Averages are merely statistical abstractions; only the individuals of which the populations are composed have reality. The ultimate conclusions of the population thinker and the typologist are precisely the opposite. For the typologist, the type (*eidos*) is real and the variation an illusion, while for the populationist the type (average) is an abstraction and only the variation is real. No two ways of looking at nature could be more different.[21]

Mayr's population thinker stressed that populations persist only to the extent that they function as reservoirs for multiple variations of any given trait. The fact that each individual in a population is genetically unique contributes to long-term population fitness by extending the ability of the population to respond to *changes* in environmental conditions. Mayr understood a population as a fundamentally "speculative" biological unit, in the sense that individuals of a population who were less fit for current conditions at the same time functioned as speculative hedges against the possibility of significant changes in the environment.

For many biologists of the twentieth century, including Mayr, the speculative dimension of biological populations was captured both theoretically and visually through the tool of a *fitness landscape*, first proposed by Sewall Wright in the 1930s and cited by Mayr in *Systematics and the Origin of Species*.[22] Wright, one of the founders of population genetics, was also a significant figure in the development of the modern synthesis of Darwinian thought, primarily through Wright's development of statistical tools and theories—for example, the theory of *genetic drift*—that helped geneticists make sense of animal and plant experiments with genes and chromosomes. In a 1932 paper, Wright suggested that each individual be represented as a distinct point in the field of all possible genetic combinations for that species. In an actual population, not all possible genetic combinations will be realized, in part because many combinations will prove biologically unviable. For all the viable genetic combinations, though, there will be a so-called adaptive or fitness peak in the field around which most members of the actual population cluster. However, as environmental conditions change, another peak may come to represent the greatest possible fitness for the population. Because a population is always spread out around a peak—or, to use a financial metaphor, is diversified—the population can, by means of sexual reproduction, move toward the new, higher fitness peak over time (see figure 1.2). The fact that the members of a population are spread across some subportion of the fitness landscape enables the population to deal not only with its present, but also, in essence, to bet on a multitude of possible futures. The tool of the fitness landscape helped biologists to envision how a Mayrian population is able to change in a quasi-learning-like fashion, in the sense that its genetic composition changes over time in response to changes in external conditions.[23]

Wright's fitness landscape also made it possible for both biologists and computer scientists to consider how the differences among individuals to which Mayr pointed could be quantified. Wright's representation of the field as a landscape and of fitness as a mountain or hill-like peak, and his emphasis on the importance of chance for changes in a population as it adapted to the fitness landscape, had the added virtue of aligning

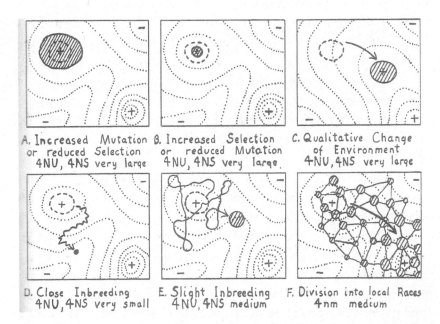

A. Increased Mutation B. Increased Selection C. Qualitative Change
   or reduced Selection   or reduced Mutation   of Environment
   4NU, 4NS very large   4NU, 4NS very large   4NU, 4NS very large

D. Close Inbreeding   E. Slight Inbreeding F. Division into local Races
   4NU, 4NS very small   4NU, 4NS medium   4nm medium

1.2 Wright's fitness landscape. *Source*: Sewall Wright, "The Roles of Mutation, Inbreeding, Crossbreeding and Selection in Evolution," in *Proceedings of the Sixth International Congress of Genetics*, ed. Donald F. Jones (Ithaca, NY: Brooklyn Botanic Garden, 1932), 361.

this understanding of population genetics with the engineering and computer science methods of *self-optimization* developed in the 1940s and 1950s, especially in the new metascience of cybernetics. Hill-climbing algorithms developed in the 1940s and 1950s, for example, employed random changes to enable a movement "upward" toward an optimal peak.[24] Wright's concept of the fitness landscape provided a bridge between evolutionary biologists and computer scientists, and by the 1960s, both biologists interested in using computers to model biological evolution and computer scientists interested in using models of biological evolution for developing optimization strategies employed Wright's concept of fitness landscapes.[25]

The speculative potential of the Mayrian population did not, of course, allow that population to "learn" in a traditional sense. The population did not, for example, represent to itself its ability to adapt to an ever-changing environment, and it could not consciously "know" anything at all. In addition, the eventual result of its adaptation was in many cases speciation, in which case the benefits of the learning enabled by the population accrued not to the original species but to the new species, which complicates further the question of what, precisely, constituted the "agent" of learning. Yet the responses of the population to its environment are remarkably analogous to a learning capacity, as computer programmers in the 1950s and 1960s were quick to note (and as we discuss further below).

At the same time that Mayr was developing his theory of populations as units of learning-like innovation in evolutionary biology, Friedrich Hayek proposed a parallel theory of *markets* as population-like units of quasi-learning. Hayek was an Austrian-born economist who became known to a large public through *The Road to Serfdom* (1944), a diatribe against centralized government economic planning.[26] Hayek was also one of the founders of the Mont Pelerin Society, an incubator for neoliberal thought and policy.[27] Hayek developed his theory of markets in the context of a more general turn in the early twentieth century, in both Europe and the US, toward centralized government planning in order to address social concerns such as health care, workers' compensation, and public works programs in cases of high unemployment. This approach intensified during the first and second world wars, as otherwise independent corporations were conscripted into the war effort, and food distribution was rationed. However, economists associated with the so-called Austrian school of economics, including Ludwig von Mises and Friedrich Hayek, argued that government planning was intrinsically flawed, at least when it impinged on economic matters.

In the 1930s and 1940s, Hayek's argument against centralized planning relied both on a claim about the intrinsic *limits* of government information gathering and a complementary claim about the ability of markets to gather and process otherwise dispersed pieces of information. On the one hand, Hayek contended that no government can gather the information it would need in order to plan economic activities—information about, say, raw materials, production costs, and consumer preferences—because this information can never be brought together at a single point. Rather, this information exists "solely as . . . dispersed bits of incomplete and frequently contradictory knowledge which all the separate individuals [of an economy] possess."[28] Hayek contended that each individual is situated in, and has the most knowledge of, their own particular "time and place," and "local conditions."[29] As a consequence, "practically every individual has some advantage over all others in that he possesses unique information of which beneficial use might be made, but of which use can be made only if the decisions depending on it are left to him or are made with his active cooperation."[30] Hayek argued that this distributed knowledge is especially important in the context of changing economic conditions, such as rising or falling production costs or changes in the availability of raw materials.[31] He argued that the only possibility of "planning" in such a state of distributed knowledge is to enable economic competition, for "competition . . . means decentralized planning by many separate persons."[32] For Hayek, "the price system" of capitalist competition functions as the mechanism by which distributed individual perspectives are brought together and by which economic problems are thus solved.[33] Or, as Hayek put it, "The whole acts as one market, not because any of its members survey the whole field, but because their limited individual fields of vision sufficiently overlap so that through many intermediaries the relevant information is communicated to all."[34]

Where centralized government planning necessarily ignores most of the diverse perspectives of individuals in favor of the specific perspectives of those in charge of planning, a competitive market synthesizes all of these limited perspectives. As in the case of Mayr's geographically isolated population, Hayek's market learns not in the sense that knowledge is gathered at a single point and made self-conscious (that is, there can be no "survey [of] the whole field") but rather in the sense that the market employs differences between individuals as a distributed means of solving problems in the context of changing conditions (and especially volatile change). The task of the economist, for Hayek, was to optimize the human use of markets. "The price system," Hayek wrote, "is just one of those formations which man has learned to use (though he is still very far from having learned to make the best use of it) after he had stumbled upon it without understanding it." Learning to make the best use of the market meant, for Hayek, actively reconfiguring as many social relations as possible *as* economic relations, for only in this way could social problems be solved. This meant understanding the market as a form of problem-solving that "has evolved without design (and even without our understanding it)" but that nevertheless enables humans "to extend the span of our utilization of resources beyond the span of the control of any one mind" and by this means enables a kind of problem-solving that occurs without "the need of conscious control."

Although there are obvious parallels between Mayr's approach to populations and Hayek's theory of the market, there are also several subtle differences (and we will return to this point in our later discussions of resilience in ecology to argue that Mayr's approach points toward positive possibilities foreclosed by Hayek's approach). Both Mayr's population and Hayek's market solve problems engendered by changing conditions: for Mayr, differences among individuals enable a population to adapt fairly quickly to a new environment, while for Hayek, differences among the individuals connected by a market enable that market to adapt quickly to changes in resources, consumer demand, methods of production, and product distribution paths. Yet for Mayr, the speculative capacity of a population—the fact that differences among members of the population serve as hedges against multiple forms of environmental change—often resulted in the transformation of that population into a *new* species. For Hayek, by contrast, the market does not transform into something else but can only be optimized. On the one hand, this seems to make the Hayekian market the analogue not of Mayr's population but rather of the forces of natural selection that operate on populations. Yet for Hayek, differences among individuals are themselves part of the "environment" of the market since what is selected for by the market are different consumer products and services, which are then used by those individuals. This ambiguity of the Hayekian market—is it the analogue of a population, the forces of natural selection, or both simultaneously?—was not, as it turned out, an impediment for this theory but rather a means by which it could link itself to the "objectivity" (and hence, virtue) of a scientific evolutionary theory of populations even while maintaining the inviolability of the market itself.

While economics had long been linked to population—Malthus, for example, was a professor of political economy—Hayek's understanding of the market as a population of unique individuals represented a new formulation of this link. Different understandings of population, evolution, and adaptation are distinguished from one another by, to paraphrase cybernetician Gregory Bateson, differences that make a difference. In the biological theory of evolution understood as enabled by differences among members of populations, the main issue is adaptation to environmental change, even if that sometimes results in the extinction of species. For Hayek's theory of the market, evolutionary concepts are still linked to change and extinction, but these now apply *only* to market participants and their products. The concept of evolution thus functions in neoliberal economics as a tool to enforce the need for competition among participants in the market, it but cannot be extended to ask the question of whether the "freedom" currently instantiated in "free" markets can itself be subject to evolution.

The more recent smartness mandate takes up Hayek's population-adaptation logic but then transforms this latter in yet a new direction. Hayek was still centrally concerned with the freedom and sovereignty of the liberal subject—a concern evident in, for example, Hayek's concern that Western nation-states were on the "road to serfdom"—and justified the market as the sole guarantor of the liberal subject's freedom. While Hayek believed that markets had originally emerged without any conscious planning, he did not believe that the *best* (i.e., optimal) form of the market would emerge spontaneously, and neoliberalism was for Hayek the project of seeking actively to construct that ideal, or best, market that could best guarantee the liberal subject's freedom. Yet by the 1970s, in the wake of global environmental and "population" crises (which we discuss below), Hayek's focus on individual "freedom" had already begun to seem a bit dated, and so the "adaptive" capacity of markets would come to be justified more primarily in terms of their ability to accommodate unanticipated crises that emerged from an always-too-complex environment.[35] Rather than serving to protect the freedom of the liberal subject, the imperative for change or adaptation amplifies and supports the intrusion of information-making markets into every realm of life and even justifies the death or extinction of older markets.

While it is important to recognize the differences between Hayek's original theory of the market and more recent smart approaches to markets, it is nevertheless the case that theories of markets tend to reduce populations to behavioral populations that signal one another via the binary actions of buying or selling (or, more fundamentally, through the digital binary of true/false). Biology, by contrast, is able to comprehend many other forms of signals and data (and we return to this in chapter 4 in our excursus on the smart forest). Our point here is not to privilege evolutionary biology. Rather, we are stressing both that there is no *one* understanding of adaptation or evolution, and that concepts of evolution and adaptation have been applied and deployed through technologies of information in multiple ways. As a consequence, concepts of adaptation or evolution do not invariably provide theoretical support for, but can also challenge, neoliberal understandings of economies.

COMPUTING, POPULATIONS, AND LEARNING

While the Malthusian and actuarial approaches to population remained largely restricted to demographic and insurance discourses for much of the twentieth century, both Mayr's population thinking and Hayek's understanding of markets were taken up almost immediately by researchers interested in a new technical object: computers. This uptake was in part the result of fortuitous timing, as both Mayr's and Hayek's theories emerged at roughly the same time as the first computers, and interest in cybernetics served to connect Mayr and Hayek to individuals within computing circles. However, this uptake was also, and more fundamentally, a consequence of an interest shared by Mayr, Hayek, and computer researchers in models of learning, and both Mayr's and Hayek's approaches provided computer researchers with models for distributed and decentralized forms of learning. We focus here on two examples: evolutionary computation and the *perceptron* model of learning.

EVOLUTIONARY COMPUTATION

As computer engineer David B. Fogel notes in his reconstruction of the series of computer engineering approaches that were eventually grouped under the rubric of *evolutionary computation*, both computer scientists interested in optimization problems and biologists interested in modeling evolution on computers began to connect evolution, populations, and problem-solving in the late 1950s and 1960s.[36] As Fogel notes, there were multiple origins for this approach within computing science, including geneticists interested in modeling natural evolutionary processes, statisticians interested in optimizing industrial production techniques, electrical and computer engineers interested in artificial intelligence, and aerospace engineers interested in fluid modeling.[37] Although many of these efforts developed in isolation from one another, many also took inspiration from cybernetics research. As a consequence, what marks many of the early articles on these topics is a shared premise of an underlying homology—or even identity—between optimization problems and natural evolution, which led to the conclusion that the latter could provide solutions for otherwise intractable engineering optimization problems.[38] Or, as Fogel put it, "Darwinian evolution is intrinsically a robust search and optimization mechanism," and "the problems that biological species have solved are typified by chaos, chance, temporality, and nonlinear interactivity"—that is, precisely those kinds of "problems that have proved to be especially intractable to classic methods of optimization."[39] When understood in this way, biological evolution became an attractive conceptual resource for computer scientists since, by modeling the basic principles of natural biological evolution within computing, they could find solutions to problems that were otherwise impossible to solve by means of classic optimization approaches.

That meant, first and foremost, creating computer analogs of natural biological populations, for many of these early researchers assumed that it was by means of populations that natural evolution enabled a kind of "searching" and "learning." In the case of genetic algorithms, for example, a researcher used a computer to search for an optimal solution for a given problem by formatting the problem as follows:

1. The problem to be addressed is defined and captured in an objective function that indicates the fitness of any potential solution.
2. A population of candidate solutions is initialized subject to certain constraints. Typically, each trial solution is coded as a vector x, termed a *chromosome*, with elements being described as *genes* and varying values at specific positions called *alleles*.[40]

For example, a design for a robot might include multiple options for the power source (e.g., 12-volt nickel-cadmium battery, 24-volt nickel-cadmium battery, 12-volt lithium-ion battery, etc.) and multiple options for the motor (5-volt step motor, 9-volt step motor, 5-volt servo motor, etc.). The fitness of a robot could be defined as the robot's range (hours) + power (watts) – weight (kilograms). Each combination of power source and motor constitutes a possible solution to this fitness problem, and two different solutions can then be "mated" with one another by randomly combining their individual elements. The genetic algorithm is then used to determine which possible combination is the most fit.[41]

Although genetic algorithms are only one kind of evolutionary computing, most approaches to evolutionary computing rely on both competition and mating between different possible solutions within a population of possible solutions. However, even as populations of discrete biological individuals served as the inspiration for evolutionary computing, the tendency to understand evolution as a method of testing "solutions" to problems undercuts any necessary connection between discrete individuals and discrete biological populations. For example, for a statistician interested in optimizing industrial production, the relevant population was composed of small variations on an industrial process rather than the human individuals who implemented that process.[42]

HAYEKIAN MARKETS AND NEURAL NETS

An explicitly Hayekian approach to population thinking was integrated within the *neural net* strand of the artificial intelligence research tradition that first emerged in the late 1960s and now serves as the basis for many contemporary *deep-learning* computational approaches (which are, in turn, key techniques for smartness). Frank Rosenblatt's work on *perceptrons* is an important example of this integration of population thinking into premises about computation and learning that structure the neural net approach. Drawing on the work of neuropsychologist Donald O. Hebb—but also, significantly, on Friedrich Hayek's *The Sensory Order* (1952)—Rosenblatt proposed in the late 1950s that learning, whether in nonhuman animals, humans, or computers, depended upon a net of neuron-like entities among which associations would be established whenever a sensory organ was triggered by external stimuli (see figure 1.3).[43]

Rosenblatt noted that within this model, learning was not a matter of comparing external stimuli to internal models or patterns but rather

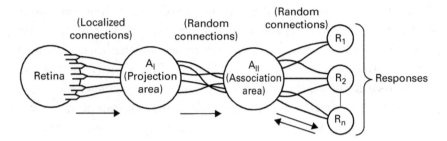

1.3 Rosenblatt's schematic drawing of a perceptron. *Source*: Frank Rosenblatt, "The Perceptron: A Probabilistic Model for Information Storage and Organization in the Brain," *Psychological Review* 65, no. 6 (1958): 389.

named the process of establishing new associations among the elements of the neural net. The key to learning for the neural net approach was exposure to a "large sample of stimuli" so that those stimuli that "are most 'similar' . . . will tend to form pathways to the same sets of responding cells."[44] As Rosenblatt stressed, this meant approaching the nature of learning "in terms of probability theory rather than symbolic logic."[45] By this he meant that learning to recognize a particular visual shape occurred when a system was exposed to a large number of instances of that shape and was initially provided with external "training" feedback; this then produced a high probability that in the presence of a new instance of that shape the same set of neurons would fire together.[46] The goal of Rosenblatt's article was to formalize mathematically how a perceptron functioned and hence how it could be instantiated in a computer device (a task that Rosenblatt had in fact completed a year earlier on an IBM 704 computer at the Cornell Aeronautical Laboratory).

While Rosenblatt mentions evolution and genetics only in passing in his article on perceptrons, his reference to Hayek helps us to excavate the implicit role of population thinking in his approach to computer learning.[47] While Hayek's *The Sensory Order* initially reads, and was explicitly presented by Hayek, as a side project largely unrelated to his economic theory, it is better understood as the necessary physiological complement of his market theory. As we noted above, Hayek's theory of the market as a population-level information processor is premised on his claim that each market participant has a unique, but also limited, perspective, and that *only* the market can bring together these "limited individual fields of vision" and hence enable economic problems to be "solved."[48] Yet within Hayek's economic theory, it is unclear whether each individual perspective is *necessarily* unique and limited, and if so, why. In "The Use of Knowledge in Society," for example, Hayek stressed differences among individuals that were explicitly the result of the demands of their occupations, such as "the shipper who earns his living from using otherwise empty or half-filled journeys of tramp-steamers, or the estate agent whose whole knowledge is almost exclusively one of temporary opportunities, or the *arbitrageur* who gains from local differences of commodity prices." These individuals, Hayek contended, "are all performing eminently useful functions based on special knowledge of circumstances of the fleeting moment not known to others."[49] In this description an individual's unique and limited perspective seems to be primarily a function of their profession combined with the imperative to maximize revenue. A captain who worries that his half-empty tramp steamer will result in personal economic loss, for example, will be attentive to the specific opportunities for filling the rest of his vessel that emerge during his travels. Yet this stress on specific professions and economic pressure did not mean that individual perspectives were always and *necessarily* unique and limited.

In *The Sensory Order*, Hayek sought a way of supporting that latter claim, and he hoped to accomplish this by describing learning itself as dependent upon what in "The Use of Knowledge in Society" he had described as an individual's "particular circumstances of time and space." Learning was dependent upon an individual's particular circumstances in the sense that although humans at birth were more or less neurologically identical, the particular neurological associations that external stimuli produced in an individual—that is, the specific instances of learning—were dependent on the particular circumstances that an individual encountered. Though this resulted in a largely shared world, each individual nevertheless remained bound to their own unique perspective, and—equally important for Hayek—there was no possibility of locating a single "correct" perspective on many matters (for example, those that relate to an individual's preferences). This made learning an intrinsically and necessarily limited accomplishment for each individual and so required some means of linking together a massive number of these individual perspectives. For Hayek, the most efficient mechanism for linking individual perspectives (rather than subordinating most individual perspectives to one individual perspective) was, of course, the market, which "acts as one market, not because any of its members survey the whole field, but because their limited individual fields of vision sufficiently overlap so that through many intermediaries the relevant information is communicated to all."[50] From this perspective, Hayek's physiological account of learning as an individual process of forging neurological connections explained why the unique individuals that were the result of such learning had to be bound together through the population-level institution of the market.

Rosenblatt's reference to Hayek's account of learning illuminates the more general role of population thinking within Rosenblatt's theory of the perceptron and its neural nets. Population thinking underwrites two different aspects of Rosenblatt's theory. It is evident, first, in Rosenblatt's attempt to clarify the perceptron model by contrasting it with what he called *monotypic models*. These latter models endeavored to stipulate in advance the precise "wiring" necessary to achieve a specific function (e.g., pattern recognition). However, as a consequence a monotypic model "is in general overdetermined, corresponding at best to a biological phenotype, rather than a species as a whole."[51] By favoring what he described as a "genotypic approach," Rosenblatt aimed at a model of learning for which "the properties of the components may be fully specified, but the organization of the network is specified only in part, by constraints and probability distributions which generate a *class* of systems rather than a specific design."[52] In other words, for Rosenblatt the goal of brain modeling was not to construct a specific wiring diagram for a specific task but rather to model a structure that accepts differences among multiple members of the same species, or class, so long as the results of the members of the class are statistically similar. Just as multiple human beings develop slightly different neuronal linkages as a consequence of their slightly different environments yet can nevertheless usually agree on whether a photograph is of a dog or a cat, multiple runs of a pattern recognition neural net would result in different associations between computer neurons, but each run would end up with statistically similar capacities for recognizing the target pattern.

Second, precisely because perceptrons require training data (as well as an agent who helps the neural net assess the training data), they can in principle be trained on what is in essence *population-level* experience.[53] Although each human individual is limited to that specific set of external stimuli to which they are in fact exposed, a computer perceptron can, by contrast, draw on databases that group the judgments and experiences of not just one individual but large populations of human individuals. As we will note in the penultimate section of this chapter, though such databases were not available in the 1950s or 1960s when Rosenblatt developed his model, they have since become standard and include, for example, the Modified National Institute of Standards and Technology data set of handwritten images, Google's Open Images, the Stanford Dogs data set (which are used to train visual pattern recognition algorithms), and the Stanford Sentiment Treebank (which is used to train preference recognition engines).[54] Algorithms trained on such databases can then link population-level experience to the specific choices that a discrete human individual makes. For example, Netflix can recommend a movie to an individual by linking an individual's past movie choices and ratings on Netflix with the movie choices and ratings of the entire population of the millions of individuals using Netflix.

Rosenblatt's reference to Hayek also underscores the extent to which this approach to learning is implicitly open to a market-driven approach. This is the case in part because neural nets are intrinsically driven by imperatives to reduce so-called cost functions. A neural net learns by adjusting to errors, and it does so by quantitatively altering "weights" assigned to each neuron. These weights, like Hayek's "prices," provide a common quantitative medium for the net as a whole (and in this sense make the neural net a market-like structure). Perhaps more significant, a perceptron can be trained with massive population-level data sets only if and when data from populations can be gathered and formatted in consistent ways, and this latter necessity creates a much deeper elective affinity between the perceptron/neural net model of learning and markets. Neural nets require consistent and quantifiable sensory input, and this is facilitated not only by omnipresent sensors but also by prices and quantified rating systems that create consistency across that data. While these kinds of market relations were absent from Rosenblatt's original formulation of the perceptron, his basic architecture implicitly pointed toward the need for population-level, consistent, and market-like mechanisms for gathering and assessing data.

THE LIMITS TO GROWTH AND WORLD SYSTEM MODELING

In the late 1960s and early 1970s, the Malthusian model of population—population as a mass incapable of learning—became linked to both computation and environmental concerns through the practices of computer-assisted population modeling developed by the authors of *The Limits to Growth*, published in 1972. Critics of this report—including Hayek—noted that the project itself was based on the assumption that researchers could know in advance all the relevant variables for modeling a world system—that is, precisely the assumption contested by Hayek's model of the market, Rosenblatt's critique of monotypic models of learning, and Mayr's valorization of population thinking. As a consequence, one important, if unintended and ironic, consequence of *The Limits to Growth* was to facilitate the expansion of the alternative approach to populations as units of learning.

The Limits to Growth emerged from a group of "scientists, educators, economists, humanists, industrialists, and national and international civil servants" who called themselves "the Club of Rome."[55] This group was responding to a more general interest in potentially imminent global crises that involved both the environment and the world population. Since the early 1960s, a series of best-selling books, including Rachel Carson's *Silent Spring* (1962), had focused on geographically extensive environmental crises. At the same time, a separate series of best sellers, such as Paul Ehrlich's *The Population Bomb* (1968), stressed the negative consequences of an ever-increasing global human population. *The Limits to Growth* linked population growth and environmental disaster by including both as variables in computer simulations of possible futures of the human population on the globe.

The members of the Club of Rome worried that faith in technological innovation as a means for solving social problems and an emphasis on the need for perpetual growth of the world economy would lead to environmental and social crises, and they were convinced that they could prove this point via computer simulations of possible futures of the human population and global environment.[56] They took an explicitly systems analysis approach to this problem, presuming that all the "basic factors that determine, and therefore, ultimately limit, growth on this planet—population, agricultural production, natural resources, industrial production, and pollution" were each mutually dependent on one another; hence, a change in one variable would have an impact on all other variables.[57] To develop a computer model able to handle this systems approach, they solicited the help of Jay Forrester, a computer science professor at the Massachusetts Institute of Technology's (MIT) Sloan School of Management.[58] Forrester had originally developed what he later labeled a "System Dynamics" approach for helping managers of industrial systems to understand and optimize commodity production.[59] This approach produced charts such as figure 1.4, which depicts links of dependency among the various elements of an industrial system. With Forrester's help, the Club of Rome group created an analogous chart for the world as a whole (see figure 1.5). By programming these interrelationships into a computer, they could alter variables and then simulate the overall effects of these changes.

The results of the Club of Rome's modeling exercise were almost invariably gloomy. In every scenario in which "world population, industrialization, [and] pollution" continued to grow, the model suggested that, within the next 100 years at most, there would be severe food, population, and economic crises (see figure 1.6).[60] However, the report's authors contended that their models also revealed the possibility of sustainable futures, in which variables such as world population, industrialization, and pollution no longer grew but instead settled into a form of "global equilibrium."[61]

The Limits to Growth had both a significant short-term and a long-term impact in multiple countries. In the US, for example, President Carter sought to ensure that this kind of global population-modeling work would be pursued in more fine-grained empirical fashion until the year 2000, and numerous groups from both the political Left and Right and from various countries developed their own world computer models.[62] Some of these models supported the basic conclusions of the Club of Rome that growth could not continue indefinitely without leading to crisis, while others—again, on both the left and right—explicitly contested

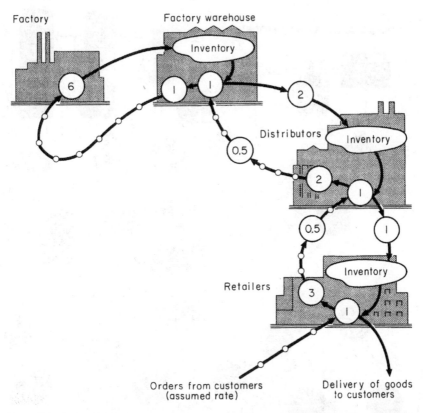

Factory   Factory warehouse

1.4 Forrester's flowchart depiction of a production and distribution system. *Source:* Jay Forrester, *Industrial Dynamics* (Cambridge, MA: MIT Press, 1961), 22.

the premise of the authors of *The Limits to Growth* that technological advances would always lag behind resource use and population growth.[63] The report was also important for the development of the one-child policy, which persisted from 1979 to 2015, in the People's Republic of China and contributed to the emphasis in USAID projects on controlling the global population (and especially the population of the third world).[64]

From the point of view of our history of smartness, *The Limits to Growth* was also important in two additional ways. First, this report established a link between populations, environmentalism, and "extraterritorial" computer modeling of global dynamics that outlasted the study's specific premises about rates of technological growth and its conclusions about the limits of growth. As the report's authors stressed in their introduction, the

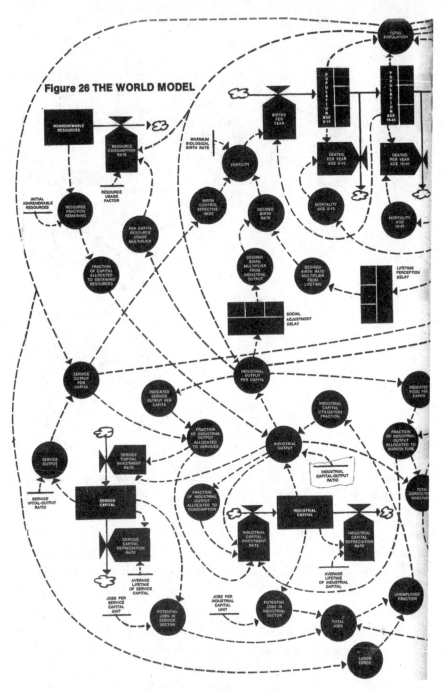

Figure 26 THE WORLD MODEL

1.5 The world model as visually depicted in *The Limits to Growth*. *Source*: Donella H. Meadows, Dennis L. Meadows, Jørgen Randers, and William W. Behrens, *The Limits to Growth: A Report for the Club of Rome's Project on the Predicament of Mankind* (New York: Universe Books, 1972), 102–103.

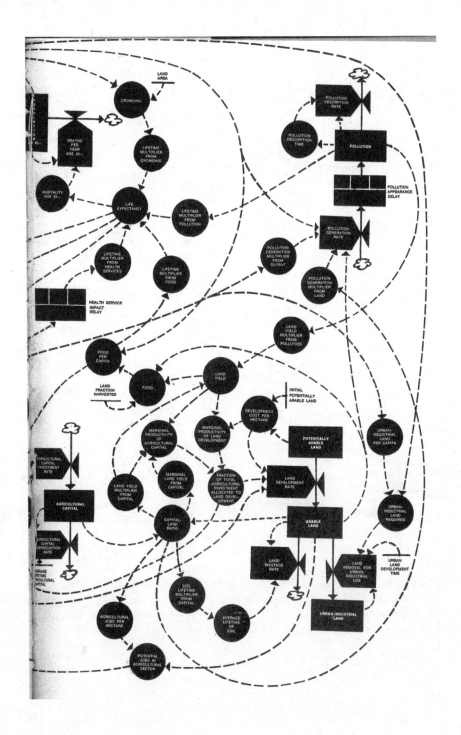

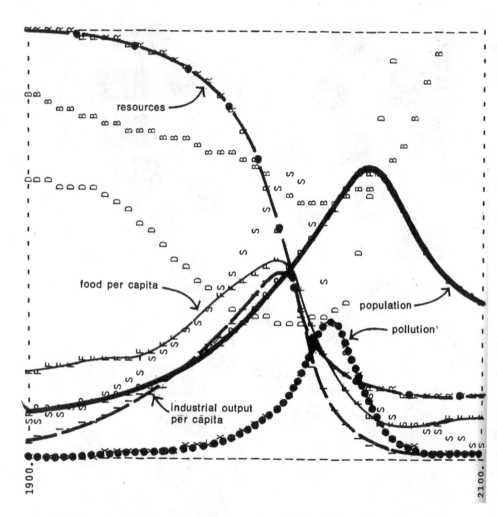

1.6 An example of what is described as a "standard" world model run from 1900 to 2100. *Source*: Donella H. Meadows, Dennis L. Meadows, Jørgen Randers, and William W. Behrens, *The Limits to Growth: A Report for the Club of Rome's Project on the Predicament of Mankind* (New York: Universe Books, 1972), 124.

problems they considered, such as global population growth and increase in pollution, lay beyond the limited frame of any particular national polity and, as a consequence, "the vast majority of policy-makers seems to be actively pursuing goals that are inconsistent" with the global perspective demanded by the Club of Rome. In the face of this problem of widespread and endemic provincialism, the authors of the report suggested that computational modeling provided the possibility of an "improved"—meaning objective, extraterritorial, and truly global—perspective. This was true, they contended, because computational modeling mimicked the process of mental "modeling" with which every human being engaged their world while at the same time optimizing that modeling by "combining the large amount of information that is already in human minds and written records with the information-processing tools that mankind's increasing knowledge has produced" (21). Although *The Limits to Growth* itself had somewhat faded from public consciousness by the end of the 1970s, the report established a premise—namely, that high-end computation could link distributed knowledge with a global population and environmental salvation—that was consistent with and linked by modeling techniques and even personnel to the global weather and climate modeling that has become a central component of our contemporary efforts to understand the effects of human behavior on the environment.[65]

The second key legacy of *The Limits to Growth* for the development of smartness was the unintentional impetus it gave to neoliberal thinkers committed to the premise that *markets* were the true sites of collective learning. Both the pessimism of *The Limits to Growth*—its conclusion that environmental and population problems could not be solved, but only exacerbated, by market-based economic growth—and its emphasis on complete expert modeling of all the variables of global "health" seem in retrospect almost tailor-made as objects of critique for the increasingly powerful neoliberal movement. Neoliberal critics argued that, precisely because the market solved apparently insoluble problems, *more* rather than less economic growth was needed, for it was only by aggressively pursuing economic growth that markets could "innovate" their way beyond the (only apparent) limits to growth described by the report.[66]

EXPERTISE AS A VECTOR FOR THE EMERGENCE OF SMARTNESS

As we noted above, concepts of risk encouraged the integration of actuarial approaches to populations into contemporary smart forms of population-based, often internet-mediated medicine. The concept of *expertise*—or rather, an attack on the concept of expertise—has been a second vector for the emergence of smart technologies. While Hayek's theory of learning was already integrated into computing applications by the 1950s, his understanding of markets as sites of distributed knowledge and learning remained a largely abstract theory even into the 1970s. It was also a theory that, despite Hayek's Nobel Prize in Economics, remained relatively outside the mainstream of economic theory and approaches. But Hayek's understanding of markets as entities that learn and his corollary attack on "experts" as frauds (since individuals could never in principle be as smart as the market) began to take on teeth in the 1980s and 1990s as computer scientists integrated an actuarial logic of populations with the premise that populations can function as learning entities. Hayek argued throughout his career that the market solved problems under changing conditions when it eliminated pretensions to centralized "expertise" and instead channeled the distributed knowledge of a population through the market's quantitative price system. Yet what Hayek called "the price system" was just one way of quantifying distributed knowledge, and smartness became possible when computer scientists recognized that variants of actuarial logic could be used to create new ways of enabling distributed learning. The development of the PageRank algorithm—and, subsequently, the founding of the company Google—in the early 1990s provides an especially clear example of the means by which increasingly widespread access to personal computers helped to link the actuarial approach to populations with the premise that populations function as learning entities and to bind both of these to attacks on traditional concepts of expertise in favor of a price-oriented market logic.

By the early 1990s, an internet structure developed two decades earlier by the US military had become the backbone of a World Wide Web that allowed many thousands of users spread across the globe to create individual web pages and, equally important, to create links to web pages created by others. By the mid-1990s, though, the enormous number of web pages available suggested to several groups the need to create semi-automated mechanisms for searching among these websites so that an individual user could find the small number of pages of interest to them. In the case of print books and periodicals, libraries had done this kind of indexing work by hand for more than 100 years, and some of the early efforts for the web involved a similar process. However, as the number of web pages moved into the millions, with thousands emerging daily, this kind of approach quickly became unfeasible, and several groups developed ways of automating indexing procedures.

One of the most successful of these efforts was developed by several Stanford University researchers, who produced what they called the "google" PageRank algorithm.[67] The PageRank system of web indexing was based on the premise that a search engine ought to judge how relevant a given page is for a particular search by taking into account the "linking" activity of actual earlier users. The PageRank developers explicitly modeled their approach on the academic citation system. For academic research communities, especially those in the natural and social sciences, the more frequently an article is cited, the more important that article is considered to be; in addition, the citations in an important article are considered more important than citations in an infrequently cited paper.[68] The developers of PageRank treated a link from one web page to another as equivalent to an academic citation. Their reasoning was that establishing a link to another page was in essence a judgment that that other page was important.[69] Hence, as the PageRank developers put it, "the importance of a Web page is an inherently subjective matter which depends on the readers [sic] interests, knowledge and attitudes," yet "there is still much that can be said objectively about the relative importance of Web pages." PageRank was designed as "a method for rating Web pages objectively and mechanically, effectively measuring the human interest and attention devoted to them."[70]

Yet even as the PageRank developers presented their use of the academic citation model for ranking all web pages as a commonsense extension of a principle that had worked well in the domain of scientific experts, the result was to reconfigure the global population of web users into a vast, globally distributed population of "experts"—or, more accurately, they eliminated the distinction between experts and laypeople. The academic citation model was premised on the principle that citation ranking must be restricted to a *specific* field of academic experts, such as cell biology. Yet for the PageRank developers there was only *one* field—namely, the one that encompassed the entire ecology of current and future web pages, no matter whether their content was a "research article about the effects of cellular phone use or driver attention" or "an advertisement for a particular cellular provider."[71] The result was that the entire world of web page creators was treated as an expert community despite the absence of any structures of baseline expert validation (for example, university tenure committees, professional societies, or journal editors) and the absence of any common method or problem. As Nicholas Carr notes, this technical approach to page ranking encouraged the premise that there was a *collective intelligence* of the web as a whole.[72]

Even as each web page was treated as a unique, expert contribution to an ecosystem of collective intelligence, the point of page *ranking* was to use a search request as the starting point for parsing these individual websites into a hierarchy that would then be broken down, in actuarial fashion, into discrete categories (e.g., "Top Ten Results" vs. "Next Ten," etc.).[73] The PageRank developers argued that they employed these actuarial groupings to enable the intelligence implicit in the linking structure to become self-reflexive—that is, to enable an *increase* of collective intelligence rather than have that intelligence drowned out by self-promoting sites (the PageRank developers claim that this method is "virtually immune to manipulation by commercial interests").[74]

One consequence of the PageRank approach to population-level learning is that its "individuals" no longer correspond to individual human beings. In the expert scientific communities from which the PageRank developers drew their inspiration, individual papers are authored—or more often coauthored—by discrete human beings, who accrue prestige (and hence further lab resources, increases in salary, etc.) to the extent that the papers they authored or coauthored are judged to be important by their specific scientific community. Yet in the global population of web pages that PageRank took as its ecosystem, there was no necessary link between a web page and an individual author. An individual, for example, might create multiple web pages oriented toward multiple interests or communities, while corporate home pages generally lacked any attributions of individual authorship. As a consequence, PageRank—unlike the academic article ranking system—was not designed to trace pages back to their expert author but simply to make claims about web pages. Thus, even as PageRank treated a link from one page to another as a "judgment" about importance, they did not need to treat these as judgments made by discrete individual human beings. Rather, they focused on an ecology of judgments without any need to determine the precise source of these judgments.

Although PageRank's analysis of a population of judgments was in one sense simply an extension of the population approach developed within the evolutionary-computing methods that we described above, the application of this approach to the realm of intentional actions undertaken by individuals is a difference that makes a difference. The detachment of judgments from individuals was in itself nothing new: neoclassical economics, for example, had long since performed essentially the same operation since it claimed to be interested only in the economic actions of buying and selling that actually occurred and was in principle agnostic about the nature or even existence of the "individuals" who performed those economic actions.[75] However, insofar as PageRank was designed to guide and hence amplify the collective intelligence of the information ecology of judgments instantiated in the world wide web, it integrated this approach to populations directly into the experiences of its millions of users.

In the 1990s and early 2000s, PageRank was by no means alone in its effort to merge widespread access to computing, attacks on traditional concepts of expertise, and automated forms of judgment as a way to amplify the collective intelligence of large populations. It was during the same period, for example, that Wikipedia developed a model of knowledge production that sought to bypass traditional encyclopedia gatekeepers, that companies such as Amazon and Netflix created recommendation engines that relied on consumer ratings of products, and that open-source and citizen science projects emerged that would purportedly break science out of its entrapment in the ivory tower of universities.[76] However, PageRank was one of the first, and in a sense most explicit, of the efforts to enable population-level learning by employing an actuarial approach to population judgments, which was in turn easily coordinated, via the sale of advertising opportunities, with the price system of the market.

CONCLUSION

From the perspective of the story we tell here, Naomi Klein's recent claim that corporate executives such as Eric Schmidt have employed the COVID-19 pandemic as a crisis that allowed them to push aside earlier popular resistance to the expansion of smart technologies is nominally correct but misses the bigger picture. Perhaps guided by her earlier account of the shock doctrine, in which a small cabal of neoliberal economists and heads of state employed crises as a way of thwarting the desire of the masses for democracy, Klein sees the same structure at play in recent events, arguing that

democracy—inconvenient public engagement in the designing of critical institutions and public spaces—was turning out to be the single greatest obstacle
to the vision Schmidt was advancing, first from his perch at the top of Google
and Alphabet and then as chair of two powerful boards advising Congress and
Department of Defense.[77]

The problem with this account, aside from its tendency to employ the figure of a cabal, is that it does not come to terms with the fact that the advocates of smartness claim to be enabling public "openness" and democracy *precisely by means of* smart-technology applications that channel the distributed knowledge of the population and allow the latter to learn, especially in states of significant and unprecedented change. The architects of the shock doctrine policies that began in the 1970s often acknowledged, sometimes even publicly, that "the market" and democracy might not be compatible; as a consequence, crises would need to be exploited (or manufactured) to introduce market reforms that might otherwise be unacceptable to the citizenry as a whole but would purport-edly result in long-term improvement and prosperity for the national population. Neoliberal advocates in this sense hewed closely to Hayek's approach, which focused solely on optimizing *the market* by enabling the distributed knowledge spread out among its population of unique partici-pants to flow freely (even if, at the same time, market logic was expanded to include essentially all social relations). While the advocates of smart-ness also seek to harness the collective intelligence enabled by large populations of unique individuals, they no longer limit themselves to *market* optimization but rather bring a population-learning approach to any form of problem or crisis that might present itself. Because of its quantitative price basis, the market is generally a part of the technical means by which smartness employs computation to link members of a population. However, because smartness is not intrinsically bound to markets, it can more plausibly argue that it enables forms of democracy in which markets have their proper place.

Klein is correct, of course, to stress the distinctions between democracy and smartness. The distributed knowledge of the population, for example, is *not* equivalent to what Klein calls democratic "public engagement in the designing of critical institutions and public spaces" since the latter requires the sort of self-conscious understanding of ends and means that the advocates of smartness claim is often impossible. At the same time, though, the distributed knowledge of the population is also not the "other" of democracy in the way that Klein's shock doctrine cabals so clearly are. Instead, automated systems judgments of smart populations are something more like the uncanny double of democratic decision-making. What thus remains largely unthought in Klein's account, and what we seek to engage more fully in the chapters that follow, is the question of the relationship of the populations of smartness to the citizens of democracy.

2

DEMO OR DIE: THE ZONES
OF SMARTNESS

Growing concerns with climate change, energy scarcity, security, and economic collapse have recently encouraged urban planners, investors, and governments to turn toward infrastructure as a site of value production and potential salvation in a world defined by catastrophes and crises. Nor is an emphasis on catastrophe and crisis limited to these actors, for the Left is equally focused on *disaster capitalism*, or an embrace of a world after humans. In short, the premise that environmental, economic, or security catastrophes will arrive—or have already arrived—is now a default assumption for many contemporary commentators. As if in response, a new paradigm of high-technology infrastructure development obsessed with "smart," "ubiquitous," or "resilient" infrastructures has emerged, and advocates assert that these infrastructures can save us from present or coming catastrophes. The smartness and resilience of these infrastructures refers to their integration of computationally and digitally managed systems, from electrical grids to building management systems, that can learn and in theory adapt by analyzing data about themselves and their human users. Humans are repositioned not simply as users of these infrastructures but also as part of the smart infrastructure itself. This vision has an agnostic relationship to crisis: whether the threat is terrorism, subprime mortgage failures, energy shortages, or hurricanes, smart infrastructures will save us.

In the introduction to this book, we briefly described the city of Songdo as an example of this vision of smart infrastructure. We begin here by noting three other examples from the recent annals of urban planning, which collectively underscore ways in which smart infrastructures are incorporated *within* parts of existing cities. While our first two examples were project proposals and the third is currently under construction, all three exemplify the logic of the demo that is central to contemporary smart infrastructure and the smartness mandate.

Our first two examples are drawn from a 2010 Museum of Modern Art (MoMA) in New York competition titled *Rising Currents*, which asked architects to offer designs for a future New York where seawater had risen as a result of climate change.[1] As we noted briefly in our Introduction to this book, the competition curators argued that climate change was underrepresented and underdiscussed in architecture, yet it was necessary to prepare our cities and habitats for this inevitable event. According to curator Barry Bergdoll, the exhibition would inspire New Yorkers to begin considering the future of their city under conditions of massive, and rapid, climatic change. Equally important, Bergdoll suggested, the economic recession that followed the 2008 worldwide financial crisis had laid the seeds (via layoffs of talented architects) for renewed "innovation."[2]

One of the most popular projects exhibited in *Rising Currents* was *Oyster-tecture* by Kate Orff/SCAPE, a project that has gone on to be funded to the tune of $60 million.[3] The project, sited off of Staten Island, proposes to grow oyster reefs as ecological barriers. The images from the project proposal show beautiful reefs, rendered with autumnal colors and watery blues, in a lagoon in which windsurfers enjoy the water. Other renderings depict happy New Yorkers eating oysters at shacks on the waterfront, combining security with pleasure. This culinary death comes with a certain irony since, in serving as an infrastructural defense against rising oceans, the oysters would slowly die off as a result of their dirty and inhospitable environments (and, in the long run, oysters globally are threatened by rising ocean acidity and temperature). This death, beautifully rendered by the architects, also embraces as aesthetically pleasing what is now assumed to be the inevitable destruction of much of New York by rising tidal waters.

Our second example from *Rising Currents*, nArchitects's *New Aqueous City*, repeats this theme of destruction made visible and aesthetically pleasing with a proposal for new zoning strategies and the literal use of bottom-up design strategies, such as placing floatation devices on the bottom of buildings and seawalls.[4] The video that accompanied their proposal depicted a storm surge and narrated, by way of the architectural intervention, its survival. As the waters rise, new real estate and agricultural opportunities emerge. When the big storm finally hits, we see individuals calmly gathering on the roof of what appears to be a fancy condominium, prepared for evacuation. As the helicopter swoops down, all is beautiful: the light is gentle, and there is no wind or rain.

The contrast between these images and the actual helicopter rescues, refugee-like encampments in professional sport arenas, and devastated environments in the case of Hurricane Katrina in New Orleans and Hurricane Sandy in New York begs the question of who would be left behind in this vision. At issue here for us, though, is not the quality or conception of these projects nor the intent of the curators. Both of these projects have great merit and open new possibilities for rethinking what constitutes urbanism, habitat, environment, and technology. Such a rethinking of the urban-rural-natural relationship is a necessary and key component to envisioning different futures for ecology (a point to which we will return at the end of both this and the next chapter). What interests us is the aesthetics, and therefore the politics, potentially encouraged in these presentations. In depicting global disasters as pleasing, even beautiful, and decontextualized, these images avoid questions about New York's degraded wetlands that the oysters are to replace; the inequities and violence built into the territory of the city through gentrification, financialization, and technology; and the inevitability of this forthcoming disaster. The images merely reenact, demo, or simulate disaster without that disaster ever really seeming to threaten the largely affluent spectators at MoMA.

Hudson Yards, our third example of resilient and smart infrastructure, emphasizes some of the differences between smart infrastructure proposals such as *Oyster-tecture* and *New Aqueous City* and the realities of the reconstruction of New York after the devastation of Hurricane Sandy in 2012. While many initiatives were proposed after Hurricane Sandy, most incorporating big-data approaches along with "preparedness" of communities to enhance resilience and improve emergency services, the actual development of Manhattan and its surrounding environs headed in a more ominous direction. One of the central developments is Hudson Yards, a former rail depot on the West Side of Manhattan being developed at a cost of (at least) $25 billion, which makes it the largest private real estate development in the United States, and one of the largest in the world (see figure 2.1).[5] As a recent newspaper article notes, "New York's $25 billion megadevelopment can withstand a superstorm or terrorist attack even if the entire city shuts down." The implicit argument is that Hudson Yards will survive even if the rest of New York does not. And indeed, in a section of Manhattan that is both the home of Google's Sidewalk Labs and touted as a smart development, the project is described as a "city within a city" or "a living room for creative workers." "Hudson yards," the website Business Insider chirps, "is more than just a collection of luxury skyscrapers. It's also a fortress that's built to survive off-grid in the event of hurricanes, floods, or terrorist attacks." The space is purportedly secure against terrorism; in part as a consequence, gaining access to the buildings or collective public spaces is not easy. A majority of the buildings are raised 40 feet above sea level, all mechanical systems are sealed by submarine doors, and the complex has autonomous microturbines that can heat and cool buildings if New York City's electrical grid fails. Hudson Yards is also green and smart: the plants are fed by rainwater, rather than city water, and it includes a vast array of smart building technologies and monitoring systems sponsored by Sidewalk Labs and a series of other companies. Not surprisingly, these systems have raised concerns about this city-within-a-city's governance and its use of data as a mode of deciding development, energy allocation, waste management, and other forms of resource management.[6]

As will be evident throughout this chapter, we share these concerns. Our primary goal here is to trace the genealogy of these new experimental territories created by smart infrastructures, whether these undergird entire cities (e.g., Songdo) or parts of cities (e.g., *Oyster-tecture, New Aqueous City*, Hudson Yards). We contend that what differentiates these contemporary smart infrastructures from earlier histories of urbanism is not the question of technology per se but rather the particular form of spatial and temporal containment and speculation engendered by the logic of experimentation, prototyping, versioning, and demo-ing. The development of smart cities follows a logic of demo-ing—that is, of constant

2.1 The Vessel, Hudson Yards, New York City, 2021. *Source*: Photo by Orit Halpern.

prototyping, testing, and updating—that never results in a finished product but instead installs infinitely replicable but always preliminary versions of these cities around the globe. In order to articulate this geneal-ogy, we focus on four premises and their associated practices, each exem-plified in the examples above:

1. Disaster and catastrophe, whether in the form of urban unrest or cli-mate change, become the key backdrop and impetus for urban plan-ning, which must now aim to provide infrastructural security against threats by means of constant innovation.

2. The source of this innovation is, in part, the inhabitants of the city, who are now understood as a population of "agents" (or consumers) rather than as a collective of citizens; the distributed activities of the urban population provide infrastructure with a capacity for learning and evolution in the face of always impending disaster, and in this sense the urban population is itself part of the infrastructure.

3. Infrastructure can capture and channel that population-based capac-ity for learning by means of ubiquitous computing, though this itself requires the capture and management of the attention of individuals by aesthetic means.

4. Because threats are constant, there is no "ideal" shape of the city or need for any memory of the urban past; instead, there can only be an interminable series of experiments and demos that enable perpetual adaption and short-term security for the city (or just parts of the city, as *New Aqueous City* and Hudson Yards make clear).

As we note below, the demo-logic of smart infrastructure is based on a form of temporal management that through its practices and discourses evacuates any historical and contextual specificity of the catastrophe. Precisely because the threat can never be fully represented or specified, all threats are dealt with in the same manner. While every catastrophe is different, the demo-logic that informs the production of smart and resilient cities purportedly need make no distinctions among these disas-ters.[7] Nor, as geographer Ash Amin notes, are these developments and this demo-logic restricted to the Global North, for smart cities offer urban planners everywhere "a way of imagining global and urban turbulence as governable."[8] In Amin's account, which focuses especially on the slum of

Annawadi in Mumbai, the fantasy of high-technology governed futurity is also one of political violence, as alternative forms of "intelligence" created through indigenous networks and tacit knowledge, as well as local needs for clean water, sewage, and community, are rejected in favor of computation and fantasies of digitization. The irony, Amin notes, is that even though smartness often fails in the context of such urban spaces as Mumbai and Lagos, it is precisely this failure that drives an ongoing belief by policy-makers that they need more of both urbanization and technology. Demo-logic eliminates, as a matter of principle, any function for memory (even "successful" past demos or experiments are not relevant for future threats) or for an aspirational or utopian future (all that one can say about the future is that it will bring unknown threats).

Our genealogy of this vision and practice of smart infrastructures has four sections. We begin by focusing on Nicholas Negroponte's MIT Architecture Machine Group (Arch Mac), especially Negroponte's vision in the early 1970s of *architecture machines*. We argue that this vision contained, *avant la lettre*, all the elements of smart cities, including an understanding of infrastructure that is in part computational, a stress on ubiquitous computing as a means for soliciting the knowledge and desires of entire populations, and an emphasis on demo-ing. We focus on Negroponte in small part because he is (perhaps inaccurately) often understood as the origin of the concept of *responsive environments*. Much more important, though, Negroponte and Arch Mac allow us to underscore in the second through fourth sections the extent to which his vision, shared by other urban planners at the time, grafted together two approaches to urban planning: on the one hand, an urban planning tradition that believed that computing could reduce or eliminate social conflicts around class, race, and political divisions (second section) and, on the other hand, a longer cybernetics approach to neural nets that implied that distinctions between what was "inside" and "outside" the net were beside the point (third and fourth sections). Sponsored by defense research funding and corporate investment, Arch Mac was directly invested in producing technologies and urban design solutions to everything from postcolonial conflict to American race warfare to the transformation of corporate research and development. Not coincidentally, Negroponte's famous imperative that summed up the high-technology start-up mentality of the 1980s and

1990s—"demo or die"—also best describes the smart mode of futurity and optimism that seeks to ward off impending disaster and that is refracted in our ongoing preparations and responses to disasters, whether these involve weather or pandemics. In an excursus, we document how Negroponte's premises have been integrated into more recent smart systems, including smart electrical grids. In the fifth section, we outline the merger of the classic urban-planning concept of the zone with the newer concept and practice of the demo, and we stress again that the link between the experimental zone and smartness is not restricted to the Global North but is also part of Global South urban planning. We close with some reflections on alternatives to this vision, stressing the importance of determining the relationship between the demo-logic of the smart city and the *demos* of democracy and the polis and thus how we might imagine growth in ways that account for history, time, and difference.

EVOLUTION, LEARNING, AND THE URBAN ENVIRONMENT

Our first reference point in the genealogy of smart infrastructure—Nicholas Negroponte's Arch Mac—may be surprising. Although trained as an architect and despite his significant influence on the world of computing-assisted architecture and media design, Negroponte is not often discussed in histories of urban planning. (He is completely absent, for example, from Stephen Graham and Simon Marvin's *Splintering Urbanism: Networked Infrastructures, Technological Mobilities and the Urban Condition*.) Yet Negroponte was among the first to outline the vision of smart infrastructures in texts such as *The Architecture Machine* (1970) and *Soft Architecture Machines* (1975). He was able to do so in large part because of his academic location at MIT, for his vision of what would come to be called smart infrastructure was a synthesis of existing MIT approaches to the study of urban processes (and problems), on the one hand, and computer learning on the other.[9]

According to architectural historian Molly Wright Steenson, Negroponte's *The Architecture Machine* became a bible for computer-aided design, precipitating the spread of computers within the field of architecture. She suggests that the text was also critical in encouraging an emergent do-it-yourself (DIY) ethos in the design fields that began in the late 1960s and

early 1970s and which continues into the present—an ethic not of institu-
tionally validated expertise but of constant experimentation, "versioning,"
and "hacking."[10] For our account, Negroponte's text is especially important
for the way in which it employed concepts of evolution and adaptation—
which, together, led to a notion of quasi-biological, computer-assisted
evolutionary learning—as justification for the insertion of machine intelligence into architecture and urban design. "I shall consider the physical
environment as an evolving organism," Negroponte asserted at the start of
the book, and "I shall consider an evolution aided by a specific class of
machines."[11] Assisting evolution by means of machines was to lead to what
Negroponte called "an environmental humanism," which meant that the
"design process, considered as evolutionary, can be presented to a machine,
also considered as evolutionary, and a mutual training, resilience, and
growth can be developed."[12]

Environmental humanism responded to what Negroponte saw as a
failing of human-only design in a modern context—namely, the difficulties humans encountered in dealing with "large-scale problems." This difficulty resulted from the tendency of human designers to employ simple,
one-size-fits-all solutions to complex problems combined with the fact
that humans are not good pattern seekers; together, these limits made it
difficult for humans to see systemic problems that resulted from changes
over time or changing contexts. Computers, by contrast, could in principle (if not necessarily in fact in 1970) "respond intelligently to the
tiny, individual, constantly changing bits of information that reflect the
identity of each urbanite as well as the coherence of the city." Negroponte outlined a vision in which every home contained a computer that
gathered "localized information."[13]

Negroponte suggested that in an urban environment in which every
home contained a machine, each urbanite could be intimately involved
with the design of their own physical environment by (in effect) conversing
with that environment about their needs. Or, to consider this another way,
each individual would implicitly be talking to the architect via a machine-
to-machine interchange.[14] Negroponte stressed that the goal of this vision
was *not* the production of uniformity but rather the enabling and social
channeling of unusual and exceptional interests, abilities, and desires:

What will remove these machines from a "Brave New World" is that they will be able to (and must) search for the exception (in desire and need), the one in a million. In other words, when the generalization matches the local desire, our omnipresent machines will not be excited. It is when the particular varies from the group preferences that our machine will react, not to thwart it but to service it.[15]

To emphasize the grandeur and infinite scope of Arch Mac's ambitions in liberating humanity from history or stagnation, major concepts in *The Architecture Machine* were introduced through a wide range of examples, such as informal architectural planning in Latin America, European hill towns, and community developments in Boston. These positive examples were often opposed to images of centrally planned, gridded cities in former colonies and/or the Global South (figure 2.2).

   Race relations within the United States played an important role in Negroponte's explanation of and justification for computer-aided urban planning. Negroponte described an experiment with tenants in Boston's underprivileged South End neighborhood, where battles over eminent domain and the resettlement of residents for commercial development and transport infrastructure were pervasive. (As we note in more detail in the next section, these political battles occurred within a context of rapid transformation of the urban economy and geography related to the rise of the finance, insurance, new real estate markets, and high-technology industries.) Negroponte and his associates recruited three African American men from the city's public-housing projects and used a computer to ask each about concerns regarding urban planning and neighborhood improvement and, specifically, what each wished urban planners and designers would take into account (figure 2.3).[16] Negroponte claimed that

the three residents had no qualms or suspicions about talking with a machine in English, about personal desires . . . instead, they immediately entered a discourse about slum land-lords, highways, schools, and the like. . . . [T]he three user-inhabitants said things to this machine they would probably not have said to another human, particularly a white planner or politician: to them the machine was not black, was not white, and surely had no prejudices.[17]

Negroponte contended that if opinion-soliciting computers were distributed in an environmental (i.e., ubiquitous) fashion, "the design task" could then focus on blending the preferences of the individual with those of the group. Machines would monitor the propensity for change of the

2.2 Examples of the abstract concepts of force, Brasilia, and vernacular architecture from Italy and Spain. *Source*: Nicholas Negroponte, *The Architecture Machine* (Cambridge, MA: MIT Press, 1970), 4.

body politic. Large central processors could interpolate and extrapolate the local commonalities by overviewing a large population of "consumer machines."[18]

In the BA thesis that he had completed in 1965 at MIT, Negroponte had been interested in the more standard question of population growth, though even in that early work he had seen differences among members of the population as a resource to be harnessed.[19] In *The Architecture Machine*, his earlier interest in the *growth* of the human population was

2.3 Public housing tenants demo-ing the Architecture Machine Group interactive urban development questionnaire from Nicholas Negroponte's *The Architecture Machine* (Cambridge, MA: MIT Press, 1970), 57.

displaced by an interest in how a population of computers could mediate individual and collective desires. The corollary to this, though, was that the human population that inhabited the city was also subtly redefined. Within Negroponte's vision, the urban population is not a community of citizens who engage with one another politically but a collective of "preference agents" whose desires are best mediated and facilitated by a population of computers. Traditional concepts of democracy require the concept of a space, whether real or virtual, within which citizens can achieve an overview of the various political claims being advanced and then make judgments about and decisions concerning the merits of those claims. Negroponte's vision of ubiquitous computing dispenses with the need and even possibility of such a space, substituting instead a population of computers that can perpetually experiment with possible resolutions to social conflicts (and, if one possible solution fails, that can try another experiment). The Greek *demos*, in other words, becomes the engineering demo, or test bed.

SPLINTERING URBANISM AND ZONES

Negroponte's situation at MIT in the 1960s and 1970s enabled this vision of environmental humanism and ubiquitous computing, and its corresponding transformation of *demos* into demo, in at least two ways. In this section we consider the more general backdrop of challenges to urban planning in the twentieth century and the ways in which this affected Boston specifically. In the next section, we consider the lineages of computation at MIT to which Negroponte's approach was both implicitly and explicitly indebted.

Negroponte's new vision of urban design was one response to what geographers Graham and Marvin have described as "splintering urbanism." As Graham and Marvin note, late nineteenth- and early twentieth-century urban planners presumed that social order could be ensured by using centrally developed urban plans and zoning ordinances to design urban infrastructures, such as road, sewage, and electrical systems. These planners assumed that treating infrastructures as public goods that reached all members of the urban and national populations equally would increase the strength and economic viability of the nation. By the 1950s and 1960s, though, this image of urban planning had come under increasing pressure. Graham and Marvin note multiple reasons for this loss of faith in the modernist vision of planning, including the neglect and decay of city infrastructures as well as the increasing privatization of infrastructure, which called into question whether the latter could truly be treated as a public good. They also stress the increasing realization that

modern urban planning . . . had neglected many voices, in its "mainstream" depiction of the modernist planner as an omniscient, benevolent (inevitably male) "hero," taming the wild chaos of the disorderly metropolis. The views of women, minority ethnic groups, indigenous people, disabled people, gay men and women, older people and children were largely ignored. Modern urban planning had often therefore ignored the essentially patriarchal, racist, disab-list, socially divisive and colonialist assumptions woven into its master plans and utopian visions, being even less concerned when such assumptions were imprinted on to cities and city life. . . . Urban highway networks, for example, which purported to deliver "access for all" and add "coherence" to cities, were often found to destroy communities, undermine interactions in places, and worsen social and gender unevenness in access to transport.[20]

These concerns were brought to a larger public by, for example, the Kerner report (1967), commissioned by President Lyndon Johnson, which sought to understand the root causes of often violent civil disobedience (a.k.a. race riots) in the late 1960s, especially in Detroit and Newark.

These problems implicit in the modernist vision of centrally planned urban order were particularly evident in Boston. Between 1950, when Boston's population peaked at 801,444, and 1970, when it dipped to 641,071, the city shed much of its manufacturing and replaced it with finance, insurance, real estate, and high technology. At the same time, Boston's suburbs saw marked growth as the population of 1.63 million increased to 2.26 million, a growth encouraged by the construction of the Massachusetts Turnpike, Route 93, and I-495. This remarkably rapid transformation was fueled by changes in the legal regulation of securities and the construction of new office spaces in older neighborhoods in the city center, starting with the Prudential Center in 1965. In 1967 the New Urban League of Greater Boston, led by Mel King, began a series of protests against the eminent domain practices of the Boston Redevelopment Authority (BRA), which was clearing mostly African American residents out of the South End in the interest of relocating highways (Route 93), building overhead rail systems, and creating shopping centers. Initial protests started at a site called Tent City (today a public-private housing project adjacent to the Copley Mall, one of the most luxurious shopping centers in the United States, as assessed by cost of retail per square foot) and were successful in stalling the BRA's plans. In 1970 Mel King began the Community Fellows Program in the Department of Urban Studies and Planning at MIT and later started the South End Technology Center @ Tent City, a still-existing partnership between MIT and the Tent City Corporation providing low-cost access and training to computer-related technology.[21]

While both the Kerner report and Mel King advocated for systemic change in helping to dismantle social structures of racism, Negroponte's distrust of centralized planning led him to advocate for a different form of intervention—namely, hyper-individuated, technologically managed, responsive environments. He was, as a consequence, not interested in modernist urban-planning devices such as zoning systems. For early twentieth-century advocates of the new urban-planning device of zoning, the key innovation of this system was its division of one geographic unit, the city, into multiple districts, or zones—for example, residential versus business zones—each with its own rules. The purpose of creating this striated space was to enable the city to grow in a rational, rather than chaotic, manner. In *Zoning* (1922), for example, Edward M. Bassett contended that zoning "encourages growth while at the same time it prevents too rapid changes,"[22] while Williams argued in *Building Regulation by Districts: The Lesson of Berlin* (1914) that the purpose of urban zoning was "in the case of a city built under the older [nonzoning] conditions to change its older parts and guide growth in its new parts so as to satisfy the new demand."[23] This vision of urban growth assumed a view of territory (including the urban inhabitants of the territory) as a neutral space that could be productively structured by means of legal regulations.

Negroponte, by contrast, envisioned the city as an environment that *evolved*, rather than simply grew. As a consequence, preestablished intentions imposed from above—whether "explicitly executed by a designer or implicitly by zoning laws"—could not, for Negroponte, mediate effectively between "local actions" and "global intents."[24] Or, as he put it in *Soft Architecture Machines*, one cannot channel the productive forces of local inhabitants by means of "strict zoning, more severe building codes, one building system imposed by law." Such measures

lack the subtlety of natural forces within which a richness is conceivable. The answer must lie in the so-called "infrastructure," a mixture of conceptual and physical structures for which we all have a different definition of interpretation. . . . For my purposes here I would like to assume an infrastructure composed of a resilient building and information technology and ask what role there might be for a machine intelligence acting as a personal interface (not translator) between this infrastructure and my ever changing needs.[25]

For Negroponte, the modernist division of cities into discrete zones, while perhaps a step in the right direction, still assumed a too static view of the future—that is, assumed that the near and medium-term future would be essentially like the present—and hence led to precisely those kinds of racial conflicts around housing, use, and infrastructure that he subtly referenced through several of the demo projects discussed in *The Architecture Machine*. Negroponte's vision of zoning, by contrast, was premised on fluidity and perpetual computer-mediated negotiation:

For example, my roof surface could serve as your terrace without inconvenience to me because it happens to be above services and functions that would be disturbed by noise. Or, I might not mind your cantilevering over my entrance, as the reduction in light would be more than compensated by the additional shelter I happened to want. While these are simpleminded examples, they reflect a kind of exchange (even bargaining) that is not possible in present contexts. They assume two parties, but this could be extended to complex and circuitous trade-offs: if A→B, B→C, C→D, . . . ,→n, n→A. We begin to see the opportunity for applying three-dimensional zoning standards and performance standards in context, a feat that I propose is manageable only with a large population of design amplifiers that could talk to each other and to host machines.[26]

It is difficult, based on Negroponte's examples in this quote, to describe this as "zoning" at all since every zone is up for perpetual renegotiation, and there is no centralized zoning authority. It is instead a mode of perpetual experimentation and change.

However exciting this approach might be from an urban planning point of view, it is hard to see how this method could address the legacies of colonial urban planning and racial conflict that Negroponte referenced in *The Architecture Machine*. Even if it were the case, as Negroponte asserted, that for African Americans the computer was "not black, was not white, and surely had no prejudices," it is not clear how those structural problems of inequitable access to housing, employment, and infrastructure to which the Kerner report had pointed could be addressed through the hyper-individualized computer-mediated zoning negotiations that Negroponte described. Instead, Negroponte's approach foreshadows a mode of market-based social relations that seem incapable of, and in principle opposed to, both democratic contestation and the use of planning to address deep structural issues such as racism.

COMPUTATION AND URBAN DYNAMICS

If problems with the modernist vision of central urban plans, zoning, and infrastructure as a public good provided one enabling context for

Negroponte's vision of ubiquitous computing, another was provided by the history of computation—especially computation applied to political and urban issues—at MIT. As historian Jennifer Light has noted, the discourse in the 1960s and 1970s of a "crisis" of US cities had been heightened by an influx of defense intellectuals leaving the analysis of nuclear strategy to apply their operations research and cybernetic methods to the burgeoning and increasingly profitable sector of urban security and development.[27] Negroponte's Arch Mac expanded on this condition. Planners at MIT generally did not address the perceived crisis of urban environments with a turn to, for example, conventional sociology. Rather, they made use of the tools of environmental psychology, communication theories, cognitive science, and computer science. For example, Kevin Lynch, who developed environmental psychology, and computer engineer Jay Forrester, who applied computer programming methods and simulations to environmental problems and to urban design issues, had been applying cybernetic approaches to human habitation at MIT since the 1950s.[28]

The question of how computing might "guide" democracy was also central to several MIT-based projects, including Simulmatics and Project Cambridge. Simulmatics was a political consultancy firm established in the late 1950s that used computerized poll data to provide political campaign advice. Founded by behavioralist-leaning professors of political science, including MIT's Ithiel de Sola Pool, Simulmatics was one of the first polling projects to include information from African American voters, and it provided its services to several campaigns in the 1960s. Simulmatics also received $250,000 to track media coverage of urban US riots so that this information could be included in the Kerner report.[29] Although Simulmatics was not financially successful and eventually folded, its basic approach was taken up within Project Cambridge, an MIT-based successor to Simulmatics. As Fenwick McKelvey notes, Project Cambridge proposed to integrate computational social science into decision-making in order to create "a powerful methodology for the behavioural sciences." When integrated into public policy, these tools were to aid "in the understanding of human interactions and in the prediction of the performance of social systems."[30] Director Douwe Yntema noted that Project Cambridge aspired to something like the self-learning, self-governing computer systems earlier envisioned by Oliver Selfridge, which we discuss further below.

If Project Cambridge's aspirations to manage social conflict by means of computation provided a more distant context for the work of Arch Mac, the urban-modeling systems of MIT computer engineer Jay Forrester were more directly relevant, for Forrester introduced new ideas of technical obsolescence into urban-planning practices. In *Urban Dynamics* (1969), a study that introduced models of computation to urban simulation (and that emerged out of Forrester's interactions with Boston mayor John F. Collins), Forrester concluded that cities must be treated as *systems* of industries, housing, and people. And hence for cities, as for all technical systems, "processes of aging cause stagnation."[31] From this perspective, the social sciences and the study of society as the foundation for urban planning were not helpful, and Forrester began with the claim that the entire social science research project had no grounding in the literature of urban studies or planning. Forrester instead advocated for the direct application of approaches from business and computer science previously used to model the growth of corporations.[32]

Forrester's research was conducted under the rubric of the Urban Systems Group, a Ford Foundation-funded initiative to apply management and computing to urban problems. The Urban Systems Group emerged at MIT in the late 1960s and was critical in linking the schools of urban planning and engineering at MIT, as well as in forming cross-disciplinary projects at the University of Michigan, Harvard University, and elsewhere. *Urban Dynamics* details efforts to use time-shared computers to run generic models of common urban revitalization programs and of urban growth, testing how different policies produce different outcomes. Forrester sorted urban processes into categories: *inputs* (which today might be labeled *stakeholders*); *valves*, which were actions that would have an impact on demographic and economic activities, such as increasing education and investing in housing; and *outputs*, or desired results (see figure 2.4).

The Urban Systems Group spent enormous energy and capital to build computer simulations with different arrangements of inputs and valves and measured the resulting urban productivity of each simulation by means of economic output and demographic indicators such as class, profession, and birth rate.[33] It is important to note that these were sui generis models grounded in policy and planning at MIT, not in studies of existing cities. What distinguished the MIT approach to urban design and planning through computing was its focus on process rather than on end points, as well as the reformulation of urban problems through the direct application of organizational management represented through the flowcharts and feedback loops of programming.

Forrester's discourse of obsolescence and management marks a critical turn in reimagining urban territories and the practices of urban planning and design. As historian Daniel Abramson notes, discourses of obsolescence are historically specific and differ from modern discussions of urbanism as a rapid, industrial, and often dehumanizing process. Cities in the 1950s and afterward were increasingly imagined as obsolete technical systems, a conception that demanded new approaches emerging from organizational management and computer-aided manufacturing.[34] Forrester thus anticipated a logic of refreshable newness—later invoked in Negroponte's "demo or die" motto—that envisioned aging and nongrowth (or stagnation) as the central cause of urban problems, race conflict, and environmental issues. The implication is that cities, like corporations, must engage in constant reinvention or else they will age and die.

Conceiving cities as corporate-computational systems distanced the imaginary of planning from ideal or utopian forms and reframed urbanization as an ongoing process of calibrating cities for constant change. Nineteenth- and early twentieth-century modern designers and urban planners, such as Ebenezer Howard, Le Corbusier, and Bauhaus designers, produced utopic and replicable *forms* of the city. Their utopian aspirations were (or were intended to be) embodied in distinctive urban shapes: the linear, body-like design of Le Corbusier's *Radiant City (la Ville radieuse)* or the linked set of concentric circles of Ebenezer Howard's "garden cities" (see figures 2.5 and 2.6).

While the early twentieth-century zoning movement was not always based on a specific shape, it too presumed an urban space in which the distinctions between zones would be clear and constant for at least a few decades before shifting to another zoning arrangement. Both Forrester's and Negroponte's approach, by contrast, rejected the principle of a stable shape for the city and, along with it, any clear notion of ideal form or utopian aspiration. They instead linked urban planning to pragmatic methods derived from histories of communication and computer science—namely, the flowchart and the demo.[35] Historian Shannon Mattern has argued that

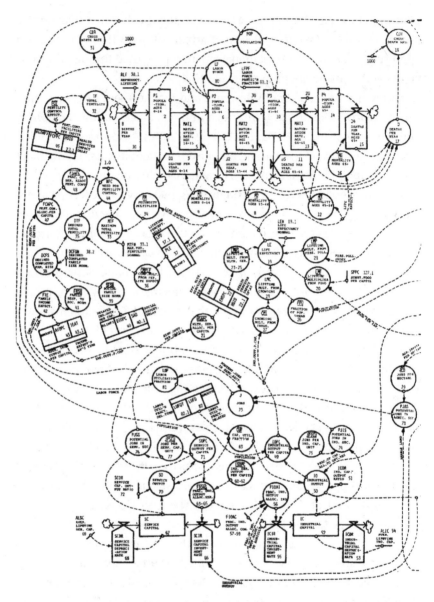

2.4 Jay Forrester's image of an urban system. *Source:* Jay W. Forrester, *Urban Dynamics* (Cambridge, MA: MIT Press, 1969).

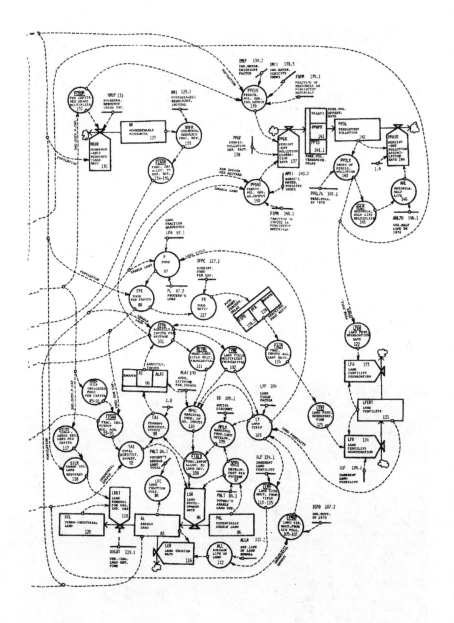

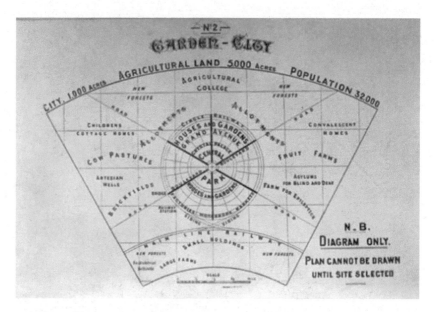

2.5 Ebenezer Howard, *To-morrow: A Peaceful Path to Real Reform* (London: Swan Sonnenschein, 1898). *Source*: Wikimedia Commons.

2.6 Le Corbusier, *The City of Tomorrow and Its Planning*, 8th ed. (New York: Dover, 2000). First published in 1929.

this type of design thinking is replicated in a conception of smartness in which the "chief preoccupations of the smart city is reflecting its own data consumption and hyper-efficient activity back to itself."[36]

ETHICAL ROBOTS, MIT ARTIFICIAL INTELLIGENCE, AND CYBERNETIC COGNITION

If Simulmatics, Project Cambridge, and Forrester's urban dynamics approach provided Negroponte with the most proximate references for integrating computation and urban design, a much longer lineage of artificial intelligence research at MIT provided him with many of his underlying assumptions. In his introduction to *The Architecture Machine*, Negroponte contended that computer-aided design demands a new form of intelligence, one no longer beholden to the human. He wrote that "computer-aided design cannot occur without machine intelligence, in fact it would be dangerous without it . . . [and that intelligence] must have a sophisticated set of sensors, effectors, and processors to view the real world directly and indirectly. Intelligence is a behavior."[37] Machine intelligence is behavioral, sensory, and decentralized; it is a smartness that is out in the world.

In his definition of architecture machines, Negroponte drew explicitly on cybernetician Warren McCulloch's concept of *ethical robots*. In McCulloch's rather idiosyncratic use of the term, "ethics" had nothing to do with morality, questions of right and wrong, or good and evil. Rather, ethics meant for McCulloch the ability to anticipate the future by learning from the past (though in a peculiar way that made it impossible and irrelevant to know the past consciously) and in this way to evolve and adapt to changing conditions. For McCulloch, ethics was equivalent to machine learning. In developing this concept of ethical robots, McCulloch was channeling a long, complicated cybernetics tradition of thinking about temporality, storage, and performativity, and this tradition undergirds the concept of the demo as an experimental practice and as an epistemology.

Cybernetics first emerged within the context of antiaircraft defense and radar research in World War II.[38] The MIT mathematician Norbert Wiener, working with neurophysiologists and doctors and influenced by Vannevar Bush's work on early computational machines, argued that human behavior could be mathematically modeled and predicted, particularly under stress, thereby articulating a new belief that both machines and humans speak the same language of mathematics.[39] In 1943, inspired by a precybernetic paper by Wiener and his colleagues and influenced by the idea that machines and minds might be considered together through the language of logic and mathematics, McCulloch and the logician Walter Pitts, working at the University of Illinois at Urbana-Champaign, decided to take the conception of the machine-like nature of human beings quite literally.[40] The resulting article, "A Logical Calculus of Ideas Immanent in Nervous Activity," published in the *Bulletin of Mathematical Biophysics*, has become one of the most commonly referenced pieces in cognitive science, philosophy, and computer science.[41]

The model of the neural net developed by McCulloch and Pitts has two characteristics fundamental to our contemporary idea of smartness. The first claim is that every neuron has a *semiotic character*—that is, may be mathematically rendered as a proposition. To support this claim, Pitts and McCulloch imagined each neuron as operating on an all-or-nothing principle when firing electrical impulses over synaptic separations. Pitts and McCulloch interpreted the fact that neurons possess action potentials and delays as equivalent to the ability to make a discrete decision. A firing, or lack thereof, affirms or denies a fact. This discrete decision (true or false; activated or not) makes neurons equivalent to logical propositions (yes/no or true/false decisions) and Turing machines.[42] The actions of neurons can be thought of as signs (true/false), and nets of neurons can be thought of as semiotic situations, or communication structures.

The second important element of the neural net model is its adoption of a strictly probabilistic and predictive temporality. Neural nets are determinate in terms of the future (they are predictive) but indeterminate in terms of the past. In the model, given a net in a particular time state (T), one can predict the future action of the net (T+1) but not which past path led to the current state. McCulloch offered as an example the model of a circular-memory neuron activating itself with its own electrical impulses. At every moment, what results as a conscious experience of memory is not the recollection of the activation of the neuron but merely an awareness that it *was* activated, at an unknown time. The firing of a signal, or the suppression of firing, can only be known as declarations of true or false—true that an impulse occurred, or false that no firing occurred—not as an interpretative statement about the context or meaning that might have motivated the firing (or lack thereof). Within neural nets, at any particular moment one cannot know *which* neuron sent the message, *when* the message was sent, or *whether* the message was the result of a new stimulus or merely a misfire. In this model the net cannot determine with any certitude whether a stimulus comes from within or from outside the circuit and whether it is a fresh input or simply a recycled "memory." Put differently, from within a net (or network) the boundary between perception and cognition, the separation between interiority and exteriority, and, more generally, the organization of causal time cannot be differentiated.

These cybernetic notions of "processing" and amnesic yet preemptive "thought" found fruition in the context of machine learning, character recognition, and research on computer vision. As we mentioned in chapter 1, in 1958 the psychologist and artificial intelligence pioneer Frank Rosenblatt, working at the Cornell Aeronautical Laboratory, published "The Perceptron: A Probabilistic Model for Information Storage and Organization in the Brain." Grounded in the McCulloch-Pitts theory of the neural net, the article argues that information processing does not take place exclusively in a centralized location like the brain. Rosenblatt hypothesized that "the images of stimuli may never really be recorded at all, and the central nervous system simply acts as an intricate switching network, where retention takes the form of new connections, or pathways."[43] There is thus no way to know if an input is new or simply recycled and no way to separate the "interior" from the "exterior" of the organism. Rosenblatt argued that rather than being concerned with the "truth" of interpreting stimuli or with older ideas of consciousness and representation in the sensory-perception-cognition system, computer scientists should focus on the route that the signal takes. It is the structure of the communication channel or the electrical or nervous circuit that creates particular responses and actions.

Assuming a nondeterministic temporality and a data-rich environment, storage in this model is not indexical, and not every stimulus is stored. What are stored are "connections or associations" and a "preference of a particular response."[44] The perceptron operates, in Rosenblatt's formulation, much like the mathematical theory of communication—that is, as a "probability theory rather than a symbolic logic."[45] The theory does not lay out a linear and Boolean representation of the process with singular and well-defined outcomes but sets up a network with certain potentials for future behavior of the machine. This model was intended to be a way to ask how intelligent computer systems might store sensory information in a manner that allows that information to affect future behavior—that is, to develop systems that could learn and respond to their environment without having implanted images of that world already within memory.

The perceptron model inspired several models of machine learning that did not rely on symbolic representation or past knowledge of a situation. In 1958, MIT computer engineer Oliver Selfridge, in communication with McCulloch and Rosenblatt, presented Pandemonium, a computer architecture for pattern recognition, at a conference on the mechanization of thought processes held at the National Physical Laboratory in London, which included many of the top figures in the nascent fields of neuroscience, computer science, and cognitive science.[46] The Pandemonium model of decentralized intelligence without symbolic processing would come to influence Arch Mac's conception of interactivity. Negroponte admitted that much of the work at Arch Mac was indebted to Selfridge, who later went on to consult for the National Security Administration on pattern-recognition software and methods.[47]

The Pandemonium architecture was—and still is—the bedrock of many machine-vision and pattern-recognition programs. It is based on the principle that instead of describing the gestalt or essence of a form and giving the system in advance an explicit definition of an object to be identified, the system could, through smaller incremental decisions, eventually find a pattern match. It is, to use a colloquial framing, a bottom-up instead of a top-down approach to software architecture. The system is composed of *demons*, programs that run quasi-autonomously in a decentralized manner as background processes, rather than falling under the direct control of the user. Each demon, composed of a neural-net cluster, or a series of logic gates, has a tightly constrained task, and there are parallel demons at different layers of the operation. For example, the first task is to recognize a very basic input—a line or corner. Demons sense the environment to find this input. They shriek when they find lines of particular shapes. The demon's shriek is equivalent to a neural fire-or-do-not-fire response (i.e., a yes/no or true/false statement). For instance, one demon deals with horizontal strokes and emits a shriek whose intensity is proportionate to how closely the data fits its search and decision-making criteria: more loudly for the letters A or T, which contain straight lines, and more quietly for O or S. At the next level up, a cognitive demon listens to the shrieks of the demon population below it and assigns a value based on the relative intensities of the shrieks—that is, how many nets were fired. Other demons simultaneously perform different tasks at this level, subjecting the letter to grids of various kinds, and the data they gather is continuously funneled up the demon hierarchy until, finally, a judgment is made by the decision-making demon in response to the part of the network that shrieked loudest. For example, if the part of the network seeking curved shapes sends out the most signals and another part seeking straight lines shrieks less, then the next level of decision-making demons weights the signal sent toward an S or an O rather than an A or a Z. Cumulatively, more signals coming from a certain region of the network linked to particular shapes indicate closer similarity to the letter in question. Within such an architecture, reading can occur without centralized thinking, and intuitive actions arise as computers make decisions in ways that appear human.

The Pandemonium model for character recognition marks a historical shift in both the forms of reason and logic being applied to, and enacted by, computational and engineering problems. As Selfridge argues, "*Pandemonium* is a model which we hope can learn to recognize patterns which have not been specified." He notes that "the basic motif behind our model is the notion of parallel processing," which he asserts is both seemingly "natural" and "easier to modify" than other linear or representation-grounded pattern-matching approaches. What is critical in these statements, made at the start of the era of machine learning, is a move toward inductive reasoning, or what McCulloch would label an epistemological experiment, rather than defining or representing problems beforehand. The principle is to allow machines to learn without the programmer having to stipulate an end point or fully represent a problem—that is, to allow machines to deal with what today we might label "fuzzy" or "wicked" problems. This is both a pragmatic engineering approach and a new epistemology for defining intelligence and learning in machines and perhaps other organisms as well. This somewhat jarring use of the term "organism" to refer to machines was a deliberate strategy on Selfridge's part. He insisted that "we are not going to apologize for a frequent use of anthropomorphic or biomorphic terminology. They seem to be useful words to describe our notions."[48] In a double displacement of both nature and ontology, Selfridge smoothed away the question of whether we are dealing with machine or human or animal intelligence by implying that working inductively without a predefined pattern will also bring us closer to the natural processes of cognition that occur in living beings. Computers and animals are both understood as information-processing machines. The Pandemonium model is, as media theorist and architect Branden Hookway argues, "predatory," in the sense that with it, a computer scientist can "colonize" any process by reconceiving it as a networked process.[49]

Both the neural net and Pandemonium models of sense perception and character recognition suggest a new cognitive-sensory paradigm grounded in making process a thing in the world, a material and technical entity amenable to algorithmic manipulation and production. Sense perception and cognition are compressed into a single channel and envisioned as both material and scalable; what applies to individual organisms can also scale to aggregates of organisms and to environments. These models posit a decentralized and networked understanding of mind and analytics. Scalability is grounded in the fundamental confusion of boundaries and the preemptive logic that is literally wired into nervous networks.

In Negroponte's epistemology, the observer is reconfigured as an agent, and the environment is rendered computationally active through an idea of intelligence as agent based, amnesic, preemptive, and environmental, lacking clear distinctions between interior and exterior. Negroponte argued that the goal of computing is "making the built environment responsive to me and to you, individually, a right I consider as important as the right to a good education." Replacing the mandate for public education, long a staple of democracy, with individualized and "meaningful" environmental responsiveness, Arch Mac embraced both a new idea of a networked observer and the self-organizing, responsive territory as the central concern of design. Cyberneticians and architects thus imagined a world of neural processing that we would now label a smart or cognizing planet where our very nerves could be directly linked to networks. They also reimagined the definition of the urban denizen: an agent, or node, and not a citizen. Smart environments have no consciousness of history and are self-organizing, grounded on decisions made through networked intelligences, rather than based on the image of an individual making reasonable choices.

IMMERSION AND THE ASPEN MOVIE MAP

In 1975 Negroponte published *Soft Architecture Machines*. In this new "soft" world, the computer disappears from sight and the user is immersed within the environment. Negroponte spoke of "omnipresent machines" that will make environments full of "responsiveness."[50] Computing, which began as a conversation and then became an experiment, had now become an environment, while the question of intelligence was now measured through the metrics of usability and engagement.

This new approach was exemplified in a structure completed in 1977, the "media room." This room had quadraphonic sound, seamless floor-to-ceiling displays, and probably several million dollars' worth of hardware.[51] It was an immersive environment, fully networking human sense and computation; it also suggested an end to architecture in the older sense of the term and its subsumption into media. The new lab was meant to be a machine for assimilating differences among people, among media, and among economies. One of the media room's first projects, and one of the first three-dimensional digitally mediated responsive environments ever built, was the Aspen Movie Map (see figure 2.7).

The map was both a way to navigate space and the outcome of a new epistemology that correlated emerging notions of computation and cognitive science with design. The Aspen Movie Map was commissioned from MIT by the Cybernetics Division of the Defense Advanced Research Projects Agency (DARPA), part of the US military. Inspired by stories of the Israeli Army's use of a mock built environment in training for the mission to rescue Israeli hostages in Entebbe, Uganda, in 1976, DARPA sought to build an entirely simulated training space. The function of the Aspen Movie Map from a military perspective was to implant geographic knowledge and cognitive maps into soldiers before they arrived at the real site in combat. But for members of Arch Mac, including project director Andrew Lippman, the main purpose of the project was not related to human memory, training, or specific geographic knowledge. Rather, it was to develop

2.7 Screenshot of the Aspen Interactive Movie Map's configuration (1978). *Source:* "The Interactive Movie Map: A Surrogate Traveling System," MIT Media Lab, YouTube, January 1981, https://www.youtube.com/watch?v=Hf6LkqgXPMU.

more interactive environments for engaging with computers and to test the emerging technologies of videodiscs and high-resolution storage and replay systems.[52] The map was built through the careful survey of a space using gyro-stabilized cameras that took an image at every foot as they covered the streets of Aspen, a high-end ski resort town in Colorado. The system employed a computer, laser discs, and a joystick or touch screen. The map could be navigated at the user's speed and choice of route. Today this model is often touted as the predecessor of first-person-shooter video games, military simulation for both battlefield training and posttraumatic stress disorder treatment, and Google Earth.[53]

In this project there are no computers to be seen. Moreover, it was not envisioned as a model; rather, it *is* Aspen. According to Michael Naimark, an artist who worked on the project, "Aspen [is a] *verb.* . . . Aspen is known for two processes, or 'verbs' relating to heritage and virtuality. One

is to 'moviemap' . . . the other is to 'Aspenize,' the process by which a fragile cultural system is disrupted by tourism and growth."[54] Naimark's point
implies that the movie map is not a representation; it is an operation,
a way to live, a way to be in the world. It is also a self-directed, trauma-
inducing event; it "Aspenizes" or disrupts ecologies. Whether disruptive or
emergent, the project was imagined by its architects, designers, and engineers not as a room, or simply an interface, but as a cultural system and
an ecology.

As one watches the film of the original demo, questions of race, urbanization, war, and society fade away within the calm embrace of interaction. What had started as game theories around military concerns
and then the simulations of artificial intelligence had now become about
total life. The video that the MIT team produced to record the project
shows an individual slowly navigating the space of Aspen. Surveying the
tranquil, affluent neighborhoods, the interface bears no resemblance to
the military purpose for which it was built. The developers took care
to ensure that footage of Aspen would be recorded at the same times
each day in order to maintain a kind of timeless, sunny consistency in
the environment.[55] The film was shot both forward and backward so the
user could navigate in reverse, place new objects into the space and move
them, and stop at sites to learn about their histories. The intent, according to Negroponte, was to have so much recorded that the experience was
"seamless."[56]

The design of the movie map's interface implicitly uses a double strategy of deferral and *méconnaisance* to prompt the user to interact with the
system. The interface seemingly resembles the familiar world, using the
conventions of documentary cinema and first-person perspective, while
the mapping system at the top of the screen resembles the usual abstract
maps for navigation. As architectural historian Felicity Scott notes, the
system was built to include animations and additional data in order to
obscure any cuts or lags in the flow of images and to produce experiences
that repress and render invisible any editorial or cinematic cuts into the
space, thus inducing a standardized temporal movement.[57] But while history, understood as homogeneous time, is available by clicking on objects
and extracting data, historicity as a discontinuous or heterogeneous flow
of time and data (Walter Benjamin's "shocks" that emerge within history)

is banished as an experience in the interest of producing an ideal of movement without interruption through the environment. (The actual system, however, often suffered lags and stoppages as a result of limitations in memory and communication between devices.) This is a temporality emanating only as a matter of user choice and self-reference, not as a feature of engaging with the environment or with others, whether these latter are human, machine, or animal.

What makes this interface seductive is that the coordinates of real and virtual cease to exist; this floating map is not stable in time and space but is generated from within the system. Users are prompted to engage with seemingly familiar things, such as houses, cars, and streets, which typify urban and suburban spaces, even as the user's visual plane is deflected from observing a space that represents a specific locale. Users interact with the interface, modulating their body and their responses to the timing and sounds of the networked space. One is simultaneously in the local and in the global: the user is experientially in a particular place while at the same time is able to see on an abstract map the relationship of that space to a broader territory. As the map slowly unfolds and the video immerses the user in media with historical distance, the self-organizing system is networked into an attentive system. The individual here is given a sense of control over the space while simultaneously being subsumed within the network. The structural politics of militarism, race, war, and security are rechanneled into interactivity in a logic that integrates users as part of a circuit in keeping with cybernetic ideals of agent-based machine learning and sensing.

The Aspen Movie Map's particular relationship to both temporality and territory foreshadows the "demo or die" adage. Negroponte clearly distinguished between the idea and practices of simulation and these new responsive architectures by calling the Aspen Movie Map a demo, as opposed to and distinct from a simulation.[58] The demo is not a simulation since it has no reality, not even a fictionalized reality, or future to which it corresponds. Rather, the demo is a technical practice analogous to a test bed in engineering or to a prototype. It is neither a representation of the real world nor a finalized reality in itself. The demonstration of the technology hangs in an anticipatory, preemptive time of anticipation for the next technical development. The demo is a particular technology for negotiating uncertain futures and for producing realities and potentials for action; it emerges from a history of machine learning and neural nets that are always in a preemptive but amnesic state. The construction of demonstrations was part of a process whereby the environment and the user would be adjusted to one another, and eventually, the discrete demo itself would be dispensed with. The culminating success of this approach was the movie map: a system that could integrate hearing, sight, and touch to create an immersive environment that is also a place and that could train users to live in this new technically generated world.

Excursus 2.1
Smart Electrical Grids

We have focused on Negroponte and MIT in small part because of the influence of both on our contemporary world of smartness and in larger part because Negroponte and the MIT computing tradition brought into focus so many elements that would go on to engender the smartness mandate. As Larry Busbea notes, Negroponte was not alone in applying the concept of responsive environments to architecture and urban design, and similar approaches were developed in architecture departments in the 1970s at the University of California, Berkeley, the University of Texas at Austin, Pennsylvania State University, the University of Utah, Cambridge University, the Hochschule für Gestaltung in Ulm, Germany, and the Institut de l'Environnement in Paris.[59] However, Negroponte's approach helps us to understand how and why this shared understanding of responsive environment could be transformed, over the next three decades, into the smartness mandate. More specifically, Negroponte's emphasis on ubiquitous computing, a dialogue between planners (e.g., architects) and machine intelligence, and his positioning of urban inhabitants as individual "consumers" underscores the potential for what we will call *recursive infolding*— that is, a blurring of the distinction between figure and ground—of both infrastructure and the concept of experimentation in the smart world.

As we noted above, Negroponte's approach led him to understand infrastructure in general terms as "a mixture of conceptual and physical structures" and, in its ideal form, as "composed of a resilient building and information technology." These mixtures would in turn enable "a machine intelligence" to "ac[t] as a personal interface (not translator) between this infrastructure and my ever changing needs."[60] While many architects and theorists would agree with Negroponte's general description of infrastructure as "a mixture of conceptual and physical structures," his specification of that abstract

Excursus 2.1 (continued)

definition—infrastructure as "composed of a resilient building and information technology"—makes it difficult to distinguish an infrastructure from its users.[61] Infrastructure has traditionally been understood as stable and continually accessible built systems that frame the activities of a population (and which are, from the perspective of their users, figuratively invisible when they function well). Street networks, for example, channel members of a population among the various parts of the city, while electrical infrastructures bring power to most members of the population. Yet in Negroponte's vision, a human population not only uses infrastructure but serves itself as an infrastructure for a mode of intelligence—namely, *machine intelligence.*

For Negroponte, machine intelligence required computing to move into the environment (with "environment" understood as that which encompasses any specific infrastructure). When computers were spread throughout the environment, the activities of humans in one kind of traditional infrastructure, such as systems of dwelling, could become an infrastructure for machine intelligence. As members of a human population employed various traditional infrastructures, their activities could be encoded into coherent, continually accessible data streams that could in turn be analyzed by machine intelligence (or what we would now call *learning algorithms*, which turn the activities of human populations into training and real-time population data sets). For Negroponte, the point of using traditional infrastructures as an infrastructure for machine intelligence was to enable the latter to illuminate ways in which traditional infrastructures could then be altered. This recursive infolding of human and machine infrastructures would enable a never-ending process of adaptation and learning.

Yet Negroponte's understanding of human individuals as consumers also opened up the possibility that "the market" would emerge as the purportedly necessary precondition for infrastructural recursive infolding and its telos of adaptation and learning. To treat users of infrastructures as consumers points, albeit implicitly, to the function of commodity *prices* as a convenient way to deal mathematically with differences among members of a population. In the context of neoliberal interest in the 1970s and 1980s in financial speculation as a method of adaptation and quasi-learning (see chapter 3), Negroponte's emphasis on consumers implied that economic markets could mediate between machine intelligence and the activities of a population within a specific infrastructure.

This possibility became a reality in the case of the smart electrical grid infrastructures that first emerged in the 1980s and 1990s. As Rebecca Slayton notes, the term "SMART Grid" was introduced in the United States in 1997 in the context of discussions of "ways of managing the reliability challenges" of recent changes to the national electricity markets. The terminology of the SMART Grid encapsulated in a single phrase the changes in electric utility computing and software, the changes in utility market strategies, and the aspirations for environmental and national security resilience that had taken place in the preceding two decades.[62]

While computing had been a key element of electrical grids since the 1950s, environmental concerns and worries about energy reliability in the 1970s encouraged a shift in the kind of computers used to manage regional energy supplies. In the 1950s, electric utilities employed large-scale centralized supervisory control and data acquisition computer systems, with the goal of using computing to minimize power loss on electrical grids. As Slayton notes, the increasing reliance of electric utilities on computing had the perhaps unintended effect of encouraging *"economic efficiency* but not necessarily *physical efficiency."*[63] Computers allowed utilities to expand the number of users, lower costs, and increase profit but only by increasing the total amount of power generated and used. In the late 1970s, both the pressure of environmental concerns—including reports such as *The Limits to Growth*, which we discuss in chapters 1 and 3—as well as concerns, in the wake of the OPEC oil embargo, about "energy security" forced utilities to reconsider how best to link computing to power generation.[64] US federal regulators encouraged smaller and more decentralized power-generation units, and utilities increasingly came to depend upon smaller microprocessor computing. This kind of computing configuration enabled *demand-side management* approaches as utilities began to employ what would become called smart meters, "i.e., devices that could measure electricity usage on an hourly basis . . . and time-of-use pricing plans to encourage users to shift power consumption from times of high demand to times of low demand."[65]

Power utilities favored this new integration of computing and power generation primarily because of its economic benefits, rather than for the ways in which it could enable the integration of new independent power generators.[66] For "researchers, policymakers, and utilities executives," the primary virtue of the microprocessor was that it "enable[d] a more competitive market structure, reducing or eliminating the need for regulators to establish market prices."[67] This more competitive market for energy was presented as a safeguard for "the environment," in the indirect sense that a more competitive market would purportedly reduce energy waste. Kurt Yeager, director of the Generation and Storage Division of the Electric Power Research Institute, claimed in 1990, for example, that "through the use of innovative microprocessor technologies, improved resource refining and increased electrification of the economy, electric utilities can gain both social and business advantages and at the same time protect the environment."[68] Negroponte had implied that learning and adaptation would result from a direct feedback loop between human populations

Excursus 2.1 (continued)

and machine intelligence. However, in the case of electrical grids, *markets* were positioned as the key mediating infrastructure necessary for learning and adaptation (including adaptation to the natural environment).

This market-based approach encouraged energy utilities to engage in increasingly complex financial relations among themselves, such as the creation of "spot" national wholesale markets for power. This in turn facilitated greater interdependency among utilities across geographic regions, which increased the likelihood of system failures. In order to address this potential for widespread electrical system failures, utility engineers in the 1990s named and began envisioning smart electrical grids. The smart grid was premised on the following worries: "As open transmission access is becoming a reality, a major concern of electric power utilities is to maintain the reliability of the grid. Increased power transfers raise concerns about steady-state overloads, increased risks of voltage collapses, and potential stability problems. Strengthening the protection and control strategies is what utilities must do to prevent a local problem from spreading to other parts of the grid."[69] This article's authors contended that these problems could be solved in part by means of "smart devices" and "smart algorithms," which enabled "local decisions based on local measurements and possibly selected remote information" via smart algorithms "that can predict collapse from local information."[70]

It remains an open question whether smart electrical grids can actually produce more environmentally sustainable power usage or—as seems to have been the case for many smart electrical grids in the US—simply enable both greater profit and a tendency toward systemic breakdown.[71] In either case, though, smart electrical grids are a prime example of the way in which an emphasis on what Negroponte called "consumer preferences" was a vector by means of which infrastructures became smart. This kind of smart infrastructure has now become a standard demand even of the political Left: the 2019 Green New Deal proposed in a US House of Representatives resolution, for example, demands "building or upgrading to energy-efficient, distributed, and 'smart' power grids, and ensuring affordable access to electricity."[72]

In the face of this history of smart electrical grids, we find productive contemporary efforts to shift the idea of "choice" away from its capture by concepts of *consumerism* and toward concepts of political agency. This opens up new understandings of what recursive infolding can mean and how the smartness of an infrastructure such as an electrical grid can be defined. New models are emerging in Indigenous calls for environmental custodianship and sovereignty, for example, that often challenge carbon-based energy sources and infrastructures. As Winona LaDuke and Deborah Cowen note,

Infrastructure is the *how* of settler colonialism. . . . The transformations of ecologies of the many into systems of circulation and accumulation to serve the few is the project of settler colonial infrastructure. . . . Yet, infrastructure is not inherently colonial—it is also necessary for transformation: a pipe can carry fresh water as well as toxic sludge. . . . We suggest that effective initiatives for justice, decolonization and planetary survival must center infrastructure in their efforts, and we highlight *alimentary infrastructure*— infrastructure that is life-giving in its design, finance, and effects.[73]

As LaDuke and Cowen suggest, recognizing colonial and extractionist inheritances of contemporary infrastructures helps us to construct new relations between people and energy that might break that colonial history, and their notion of alimentary infrastructure is an intriguing way of reconceptualizing the recursive infolding of infrastructures that is fundamental to the concept of smartness.

LaDuke and Cowen exemplify alimentary infrastructure with contemporary efforts to establish Indigenous energy "autonomy" in Canada and the United States. In Canada, 257 of the 292 remote Indigenous communities depend on the microgeneration of electricity, often through costly and carbon energy-guzzling diesel generators.[74] However, LaDuke and Cowen narrate the story of the Navajo Nation's refusal to purchase an aging power-generation station that would require the continued extraction of coal from the lands and their turn instead to the Kayenta Solar Project. This project today powers 36,000 Diné homes and businesses across the Navajo Nation and contracts to supply energy to Phoenix and sites in Utah via a network of renewable energy tax credits and specialized products. The project was developed by the Navajo Tribal Utility Authority (NTUA) and constructed by the Spanish renewable energy firm Isolux Corsán Group and Swinerton Renewables. The development also benefited from tax credits and venture capital from organizations such as Massachusetts-based ATN International, which supported the development and construction of the NTUA's Choice Wireless, a rural broadband network also owned by the Navajo Nation.[75] According to its advocates, the grid is smart in the sense that it is highly networked, it employs state-of-the-art photovoltaic cell technology, it serves as a beacon for the Indigenous production of renewables, and it helps to end reliance on carbon energy extraction.[76]

The absence of infrastructure is also often a form of violence. In this case, the 55-megawatt installation rectifies historical wrongs even as it highlights ways in which coloniality haunts infrastructures in the present. While this project was developed through complex funding instruments and tax credits in liaison with a range of corporate and global actors, it is also a demonstration of how a new understanding of the smartness of an electrical grid can transform political relations. It does so in part by creating new forms of value and economy,

Excursus 2.1 (continued)

in the sense that this solar economy replaces the older extraction economies of coal mines and coal energy generators (which are almost depleted, in any case).[77] While those earlier extraction economies were grounded on dispossession and on environmental degradation, these new smart infrastructures promise new modes of social relations. As the manager of the project asserted, "this is what sovereignty is really about, where our people can work and provide wealth and sustain themselves." The president of the Nation, Jonathan Nez, presented the installation as an instantiation of *shánídíín*, or "a healing ray of light."[78] While it remains to be seen how this new economy will develop, we see potential for this new kind of smart infrastructure to serve new needs and create new forms of life so long as it remains attached to critical histories and different forms of value and imagination of the future.

ZONES, EXPERIMENTS, AND DEMOS

We can further deepen Negroponte's concept of the demo by articulating its relationship to both concepts of experimentation and to the modernist concept and practice of (urban) zoning. As in the case of recent advocacy in the field of geography for *geographies of experiment*, Negroponte understood urban space neither as a given nor as a neutral container within which otherwise independent human relations took place but rather as something *produced* through the interactions among humans, technologies, and geographical features (to name just a few of the relevant elements). For Negroponte, urban space had always been experimental, but he sought to use environmental computing as a way to transform implicit into explicit experimentation. However, the key role of computing in this shift from implicit to explicit experimentation meant, for Negroponte, that one must dispense with both the premise of an external observer who could assess the results of experiments and the notion of human-directed planning more generally.

The concept of experimentation was central to *The Architecture Machine* and *Soft Architecture Machines*. Negroponte claimed that *Soft Architecture Machines* "reports on a series of experiments conducted by the Architecture Machine Group at The Massachusetts Institute of Technology from 1968 through 1972," and he retrospectively described *The Architecture Machine* as "written as an epilogue to three years of experimentation that yielded both technical achievements and philosophical setbacks."[79] Yet Negroponte did not believe that one should *first* engage in experimentation in the lab or studio, with the goal of determining what works and what does not, and *then* incorporate the results of successful experiments into actual urban planning. For Negroponte, that traditional approach to experimentation presumed that the creative capacities of the urban population could be modeled or simulated in the lab. This was precisely what he felt could *not* occur. Instead, only a population of computers extended throughout the human population could mediate that potential creativity. Experimentation, in other words, had to move out of the lab and into the very fabric of the city.

Negroponte's ideal of the city as a site of urban experimentation may sound like a reprise of the classic early twentieth-century account of "the city as a social laboratory" developed by University of Chicago sociologist Robert E. Park and his colleagues.[80] Park and his collaborators understood the city as a collection of "natural areas," each of which "comes into existence without design, and performs a function, though the function, as in the case of the slum, may be contrary to anybody's desire."[81] The point of sociological research about the city, however, was "to determin[e] with more definiteness the conditions under which social experiments are actually being carried on." That information would in turn inform legislation and so enable urban and planning "reforms [to be] conducted by experts rather than by amateurs." Sociological research on the city would, in other words, transform implicit into explicit urban experimentation and so would "make the city in some more real sense than it has been hitherto a social laboratory."[82]

Yet Negroponte turned to computing precisely because he rejected the concept of the expert that underwrites Park's vision of the city as a social laboratory. For Negroponte, the information-gathering capacities of the sociologist-expert would always be too limited. Equally important, the legislative expertise to which Park's sociological studies were directed was, for Negroponte, an illusion, for Negroponte's mode of experimentation had no place for a *human* observer point from which the success or failure of experiments could be assessed. Negroponte instead aimed at a never-ending series of demos, each of which aims to take up elements of earlier demos and adjusts them for new contexts and problems, and each of which is primarily guided by machine intelligence. There was, for Negroponte—and in contrast to Park—no "outside" of the experiment, which meant that there was no external vantage point from which the city could be "planned."

Negroponte's approach to experimentation and demo-ing explains his rejection of key modernist strategies for urban design, such as the legal subdivision of the city via zoning. Urban zoning was an early twentieth-century innovation that, much in the style of Park's sense of social experimentation, drew on an older sense of zones as natural spaces but reconceived these spaces from the point of view of expert planning. "Zone" had served as a technical term within Western philosophy and the sciences for several thousand years, referring to a geographic area that had its own sui generis rules and principles. Ancient Greek philosophers, for example, used the term—which meant "belt" or "girdle"—to divide the earth into five bands: two frigid zones, one at the top and one at the bottom of the earth; two temperate zones located directly below the frigid zones; and one torrid zone, which lay between the temperate zones.[83] This basic sense of zone as a geographic band with its own specific conditions and rules underpinned the expansion of this term into various late nineteenth- and early twentieth-century discourses: within ecological discourse, for example, more restricted regions, such as mountains or seas, were divided into further zones, and the world was divided into different time zones.[84] While early twentieth-century urban planners in cities such as Boston, Los Angeles, and New York drew on this older sense of zones to make legal distinctions between business and residential zones, their key innovation was the division of *one* geographic unit, the city, into multiple districts or zones, each with its own rules.[85] This new understanding of zones—as necessarily linked to one another and as enabling planned growth of the city—also set the paradigm for the emergence of "free trade zones" in the 1930s, which constituted states of exception within the otherwise homogeneous national spaces.[86] From this new perspective on zoning, what enabled growth, whether for a city or a nation, was the differential relationships of zones to one another, and the astute employment of those differentials by urban planning experts in their designs.

For Negroponte, though, "complexity is not designed, it evolves."[87] The problem with urban zoning plans, from this perspective, is that a zoning system "assum[es], like any law-arbitrating system, the ability to exercise rules in context (which is not easy)." As a consequence, Negroponte suggested, urban zoning plans in practice ended up being modified perpetually by means of variances (that is, legal local exceptions within a specific zone).[88] Negroponte was much more interested in the traditional, "vernacular" construction of urban spaces, which he claimed liberated perpetual experimentation, and he sought to employ ubiquitous computing and demos as a means of regaining this space of experimentation within modern urban structures and spaces.

Perhaps propelled in part by the emphasis on growth and change that early twentieth-century advocates used to justify urban zoning plans, the concept of zone itself underwent something of a shift during the second half of the twentieth century. Since the 1950s, scholars in multiple discourses have been interested in the forms of emergence that occur in the zones *between* zones. Ecologists, for example, became interested in what they described as "buffer," "transition," or "junction" zones that emerge between more regular ecological zones, while scholars in the social sciences and humanities have found themselves attracted to "contact zones," "trading zones," and "critical zones."[89] This recursive infolding of the concept of zone to refer to the space between zones is a way of focusing attention away from the stable structures within zones and toward the question of where and how zones *change* (and enable change).

From this perspective, Negroponte's interest in the constant renegotiation of what we might call microzones—as well as his basic concept of urban inhabitants as *consumer machines*, which already collapses the distinction between residence and business—took this new twentieth-century logic of zones to its extreme, for it aimed at a situation of completely fluid zones in which constant experimentation is the norm. He took very seriously, in essence, the geographer's premise that space is constructed and that every construction of space is an experimental reconfiguration of the relations among humans, technologies, and environments and he sought to use computers to maximize the number of possibilities for new modes of experimental space.

Negroponte's emphasis on experimentation and demo-ing also brings back to the center a zonal function that was in earlier colonial enterprises restricted to the colony or periphery. As we noted above, Negroponte's conception of the demo emerged from a post–World War II American conception that sought to negotiate emerging decolonial and racial orders by way of experiments or "tests." As early as 1943, and as part of American war efforts, the modernist architect Richard Neutra described as a "Planetary

Test" (as well as "experiment" and "laboratory") a project in Puerto Rico
that sought to reconceptualize temperate climate zones as ideal sites for
human habitation and as the locations within which to test forms of
habitat for the future.[90] And as architectural historian Ginger Nolan has
insightfully argued, Negroponte's faith in computation and DIY design
drew on legacies of colonial enterprises, such as Marshall McLuhan's concept of the "Global Village," which emerged from the context of Kenyan
decolonization.[91] All of this underscores a point also made in histories of
epidemiology, clinical trials, and "green" crops: namely, that racialized
and Global South territories have often served not only as the recipients
of technology but also as the sites of production of new forms of biopolitical techniques.[92] In the older colonial vision, the colony serves as
a zonal testing ground for techniques that can later be brought to the
colonial center.

Yet even as contemporary discourses of testing underscore the extent
to which newer digital infrastructures continue to be haunted by these
colonial and racial legacies, smartness is a contemporary discourse, which
also involves a new form of governmentality—one inherited from, but
not the same as, the earlier colonial orders. The colony-metropole distinction is no longer determining, for the smartness mandate has been
propagated across the planet, from greenfield cities in China to projects
like the massive 100 Smart City Policies advanced by the Indian government since 2015. India explicitly markets its smart city initiative, valued
at around USD $27.6 billion, as promoting the innovation and "incu-
bation" of technologies and has linked the funding to the support of
start-ups and the collection of venture capital. This emphasis on smart
cities, along with the Aadhaar digital identification system, makes India
among the leaders in creating, and testing, massive data infrastructures
for managing populations. As media theorist Nishant Shah notes, these
new smart and digital systems create entirely novel ideas of individuals
and populations.[93] The Indian example also illuminates especially clearly
our point about the link between zones and experimentation since every
Indian smart city is able—and in a sense, is encouraged—to have a different form and understanding of smartness.[94]

The smartness mandate aims, in essence, to replace the concepts and
practices of zones understood as rule governed (and hence planned,

whether by a divine or human intelligence) with a concept of experimental zones as generative mixtures of different and shifting rules. Ideally, from the perspective of the smartness mandate, there would be only one Zone, which is itself made up of constantly shifting and recombining subzones.[95] Or, eliminating the idea of zones entirely, which was Negroponte's approach, there is a planetary space of experimentation in which smart infrastructures run constant and multiple demos and experiments with their populations, which then provide feedback for the next set of demos and experiments. This is a vision devoid of any modernist nostalgia for durable structure and form. It is also a vision that rejects the notion of modernity itself: smartness is simultaneously a hypertechnological future that nevertheless returns to traditional vernacular construction and a nonteleological embrace of biological evolution.

CONCLUSION: ALTERNATIVE EXPERIMENTS

The smartness model of constant repetitive futurity, technical failure, and demo-ing has enormous implications for how we design buildings, cities, and technology. It dominates our imaginaries and affects, as well as our responses to everything from global warming to terrorism. We live life in demonstration mode; media do not simulate or separate us from a "real" world but rather create worlds and futures. The essential question, then, is how to encounter this demo, or test bed, that has now become our world.

If we consider the elements that collectively constitute the genealogy of smart infrastructure and its commitment to a "demo or die" logic, the difficulty of our task becomes clear. The following elements, each in principle independent of the others, are now collectively bound together in the form of the demo:

- An approach to computer learning, and cognition more generally, in which there is no verifiable "outside" and no need for—or often even capacity to access—the past *as* past or memory.
- This vision of learning and cognition is mapped onto an understanding of biological evolution as a competition-driven learning process.
- The importance of evolution as the frame for imagining change over time emphasizes the creative capacities of populations, which in turn

requires a reimagination of human individuals not as citizens but rather as what Negroponte called consumer machines.

- Ubiquitous computing is the means by which those individual desires are collectively and algorithmically mediated; hence the importance of the management of attention, often through aesthetic means (e.g., the Aspen Movie Map).
- Threat and disaster, the specific nature of which cannot in principle be specified but can be exemplified by everything from urban rioting to climate change, are that around which the design of urban environments must be centered.
- There is a consequent movement away from ideal or utopian forms for the city and toward a logic of constant evolution and demo-ing, which often results in the production of "secure" sections of the city (e.g., Hudson Yards) at the expense of the rest of the city.

Although the limitations of each of these elements are clear, they have become so tightly bound to one another that simple tweaks of any component are likely to have little overall effect. It is difficult to imagine, for example, simply adding a citizen component to the consumer-machine logic that drives contemporary ubiquitous computing, as that would only further transform citizenship into a series of consumer-like preferences (e.g., likes/dislikes). Nor, in the face of global waters that are indeed rising, does it seem possible to reject either evolution or security as desirable goals. Yet it is equally difficult to imagine, in a context of perennially exhausted urban and national government budgets, that such projects could be expanded to include all of the urban population.

At the same time, rethinking these elements can begin to alter the internal connections among them. This can mean, for example:

Making distinctions among kinds of threats: As we noted above, much of the allure of Negroponte's approach to urban design was its promise to use distributed computing to resolve problems as varied as urban racial tensions, the violent legacies of colonialism, and the practical difficulties of urban zoning systems. Yet the micronegotiations among urban inhabitants that Negroponte imagined, and that have in fact become the default mode of the smartness mandate, seem incapable of resolving (and often exacerbate) problems such as structural racism and the inequitable relationships to health, housing, and jobs that it engenders. Structural racism may be a threat, but it is a threat of a different kind than global warming (even if the two are related to one another). To make this distinction is to acknowledge that structural racism will almost surely require politics and planning in the "old" sense—that is, groups staking claims to serve as the basis for a consciously created plan for the future—rather than computer-mediated micronegotiations among isolated individuals. The recent occupation of urban space by Black Lives Matter protesters is one example of this more traditional mode of politics. The related use of the photographic capacities of smartphones to alter the aesthetics of urban change are another: instead of the seamless, sunny images of urban harmony promoted by Negroponte's Aspen Movie Map and the *Rising Currents* projects, we instead have pixelated, shaky images of police violence. Employing aesthetic distinctions (and distinctions in the way aesthetics seeks to manage attention) is already an important first step toward making distinctions among various types of threats.

Creating models of evolution and biological change in which the past and history are not only relevant and accessible but central: Though Negroponte sought to convince his readers that their options were "demo or die" death is not the only option to perpetual demo-ing. It behooves us to examine other moments in which urban planners drew on biologically oriented concepts and approaches to address disasters but imagined change and memory differently. The "Metabolists" affiliated with (although also critical of) Japanese architect Arata Isozaki's alternative vision of the future of a high-technology Asian city is one such example.

Isozaki was one of the leading architects in post–World War II Japan. He first worked at Tange Lab, an architectural research and design laboratory at the University of Tokyo led by Japanese architect Tange Kenzō and his associate "Metabolist" architects. The lab was critical in creating what scholar Yuriko Furuhata has called *future engineering*. As part of this vision of importing new forms for making futures, the lab was active in the Japanese chapter of the Club of Rome, fostering institutional connections to MIT and Harvard and keeping close ties to American architects and designers such as Buckminster Fuller. Tange Lab led the cybernetic turn of Japanese architecture amidst the logistical and computational shifts in the 1950s and 1960s. Many of Tange Lab's techno-futuristic designs that

2.8 Arata Isozaki, *City in the Air*, 1962. *Source*: Courtesy of Arata Isozaki & Associates, 2021.

foregrounded the optimization of the "metabolic" processes of energy circulation, infrastructural growth, biopolitical governance, and economic prosperity of metropolises anticipated the current utopian investment in self-regulating, sentient, smart cities.[96]

Isozaki emerged from this milieux but also separated from it. As Japan was recovering from American air raids and atomic weapons, it became important to him to contemplate possible relationships with this history of destruction. He felt that architecture must be inspired by metabolic processes of life, but also by death and destruction. His vision was a corrective to the solely progressive concept of life put forth by the Metabolists.[97]

Exemplary of this genuinely metabolic comprehension, Isozaki's *City in the Air* was a project of building capsules that were to be floating and modular, moving, allowing real-time responses to the needs of its residents (see figure 2.8). However, the future of this structure was constituted through hanging buildings with computerized control that hovered over the past— that is, the remains of Tokyo—rather than destroying all memory of the past and its traumas. Tokyo was to remain partially unrebuilt, still scarred and mostly destroyed, a constant rent in the fabric of time. These multiple temporalities were to operate simultaneously, neither returning to the past nor falling into an eternal present. Neither seeking to forget Japan's own militarism and obsession with technical death nor attempting to

avoid technology—that is, neither smart nor dumb, nor forcing a decision between the future or the past—these structures were to emphasize
and attempt to work through notions of time and change: metabolism as
architecture.

The *City in the Air*'s megastructure thus offered a solution that bridged
the dreams of developers at that time. On the one hand, it aimed at
homogeneous, replicable construction and in this way anticipated the
Hudson Yards and Songdos of the world. On the other hand, it actively
embraced a nostalgia for a return to a prewar way of life, to empire, and
to the village-like structures and small paper homes of Tokyo. It was to be
a structure that permitted the vagaries of time and understood that this
form of smartness, which grows and changes with the metabolic needs of
its residents, must therefore evolve and perhaps, eventually, go extinct.

As an architectural movement, Metabolism has its own problems, and
our point is not to present it as a contemporary alternative model of smart
infrastructure and cities. Rather, its importance for us lies in its reworking
of biological approaches to cities and architecture through its emphasis
on metabolism rather than Darwinian survival of the fittest. We might
consider, for example, its relationship to the "alimentary" infrastructures
we noted in our excursus in this chapter. Thinking with metabolism as
a guiding concept requires us to think the future differently. Metabolic
processes are mechanisms for assimilating the outside into a living body,
for producing growth and change, and for eliminating wastes from the
body. Metabolism is about simultaneous growth and decomposition. At
the same time, metabolism is not the other or opposite of Darwinian
evolution: without metabolism, there would be no evolution. (The question of whether the converse is true is more complicated, as nonsexually
reproducing living beings metabolize but bear a very complex relationship to evolution.)

Beginning to think about computer-mediated learning in relationship
to both evolution and metabolism can open up a discussion about what
constitutes management and control and how we can engage temporal
multiplicity, whether this means allowing weeds to grow as a strategy for
green cities or engaging multiple aesthetics, and not just those of space-
ships and sleek glass towers. The future of politics therefore demands
that we imagine alternative futures, which also means splitting with the

techno-futurist, and modernist, aesthetics borrowed from science fiction or from urban planning, design, and technology. We must move past questions of survival to ask not how we *must* survive the present but how we would *like* to live in the future. While this chapter is hardly a solution, it suggests the importance of thinking about a history of our responsive environments in terms of difference rather than homogeneity. We must recognize that there is still critical work to do—in architecture and in scholarship—to envision alternative images of the future and different forms of time and experience that do not operate at the scale and tempo of our eternally preemptive, affective, nervous networks—to demo without death.

3

DERIVATION, OPTIMIZATION, AND SMARTNESS

As we noted in the previous two chapters, smartness presumes that although the world cannot be known completely or objectively, one can nevertheless produce useful knowledge by means of distributed populations, provided that the members of such populations are linked computationally and algorithmically in order to produce learning. The learning population for one specific problem may differ from the learning population for another problem, and there is no guarantee that the members of a specific learning population will correspond to the human population within a specific political jurisdiction, such as a city or state. This in turn encourages the development, expansion, and linkage of experimental zones and their associated sensing technologies, which make it possible to shift easily from one learning population to another, depending on the problem at hand.

In this chapter we map the genealogy of the method by which smartness produces knowledge and learning by means of populations and experimental zones. A key premise of smartness—namely, that the world is always significantly more complex than the systems we can use to respond to it—emerged from a revaluation of the concepts of noise and waste in the 1960s and 1970s. Rather than assuming, as had been the case for Cold War cyberneticians and information engineers, that noise and waste could and should be reduced as much as possible via techniques of *optimization*, early advocates of smartness understood noise and waste as not only ineradicable but also, and equally important, sites of future value. However, noise and waste could be sources of value only when their landscapes were exhaustively surveyed via what we will call *extractive* technologies capable of sifting through vast amounts of noise, waste, and the detritus of past use by populations in order to locate small increments of changing future values on which one could bet. Precisely because extractive technologies are data intensive—that is, require massive surveys of environments and populations—they must necessarily be supported economically by *derivative* financial technologies that can ensure that small bits of value, when located, can become significant sources of future value. Smart learning requires, in other words, a linkage between extraction and derivatives. While this linkage does not eliminate the drive for optimization, it renders the latter always provisional and partial.

We begin our account of the genealogy of this new logic of extraction and derivation with the development of a new understanding of financial derivatives in the 1970s, which we exemplify with the Black-Scholes option pricing model. This tool exemplifies the revaluation of noise and also helped to inaugurate the contemporary mode of financialization, which shares relatively little with past modes of financialization. In the second section, we note that though the Black-Scholes option pricing model may seem esoteric and limited to the realm of finance, its premises about noise and value are shared by a wide variety of contemporary data-intensive models for producing value, including instances of *platform capitalism* (for example, companies such as Airbnb and Uber), brain productivity research, and the health-oriented biobanks that we discussed briefly in chapter 1. As in the case of the Black-Scholes option pricing model, each of these examples revolves around leveraging noise or waste into a source of value. In the third section, we return to the immediate post–World War II period to document the elements that enabled this new model of extraction and derivation. While Black and Scholes's emphasis on noise and limits on economic agency resonated with a general sense of geopolitical instability and uncertainty in the 1960s and 1970s, their technical "solution" to this sense of instability drew on premises about psychology and management developed earlier by researchers such as psychologist Donald Hebb and management theorist Herbert Simon.

Both Hebb and Simon contributed to a new model of the economic agent, and we link the latter to a key shift in understandings of the nature of intelligence, from the Cold War "rational" (albeit paranoid) agent to the distributed form of agency characteristic of smartness.

In a first excursus, we consider an extended example—smart gold mining and the financial markets to which it is tied—to underscore that our terms "extraction" and "derivation" are not intended as simply metaphors, and how tightly linked extraction and derivation have become in our present. In a second excursus, we emphasize the importance of this history of derivation and extraction for questions of racial justice while at the same time noting that the smartness mandate differs in key respects from earlier modes of financing and derivation. In our chapter's conclusion, we consider alternatives to this now dominant model of extraction and derivation. Must financial markets bet on carbon or mineral extraction, or could they bet on recycling or renewables? Is increased consumption inevitable, or does the rise of global urbanization produce opportunities for rethinking urban life, increasing diversity and interaction between different types of people and lifestyles, and decreasing energy consumption through new transport and densification regimes? Particularly in the midst of the COVID-19 pandemic, in which the automatic response is dedensification, privatization, and segregation, it behooves us to examine seriously what we value and how we derive value in our present.

NOISE TRADING: THE BLACK-SCHOLES MODEL

We employ the concepts of extraction and derivation in a broad sense— that is, as extending beyond the references to mineral extraction and financial derivatives to which those terms might seem to point. With that said, though, we begin with an example of a significant change in practices surrounding the pricing of financial derivatives in the 1970s, for the development of the Black-Scholes option pricing model in that period illuminates a new way of understanding, and seeking to manage through new financial tools, the role of "noise" in markets. This financial example thus helps us to locate a key premise that would come to inform the combination of extraction and derivation practices now central to the logic of smartness.

In 1986, mathematician and economist Fischer Black summarized this new understanding of the connection between noise and markets. He contended that

The effects of noise on the world, and on our views of the world, are profound. Noise in the sense of a large number of small events is often a causal factor much more powerful than a small number of large events can be. Noise makes trading in financial markets possible, and thus allows us to observe prices for financial assets. Noise causes markets to be somewhat inefficient, but often prevents us from taking advantage of inefficiencies. Noise in the form of uncertainty about future tastes and technology by sector causes business cycles, and makes them highly resistant to improvement through government intervention. Noise in the form of expectations that need not follow rational rules causes inflation to be what it is, at least in the absence of a gold standard or fixed exchange rates. Noise in the form of uncertainty about what relative prices would be with other exchange rates makes us think incorrectly that changes in exchange rates or inflation rates cause changes in trade or investment flows or economic activity. Most generally, noise makes it very difficult to test either practical or academic theories about the way that financial or economic markets work. We are forced to act largely in the dark.[1]

In the 1980s, Black's suggestion was still counterintuitive, and he made these claims in part to explain (and justify) why a series of new financial instruments that had appeared in the 1970s—some of which Black himself had helped to create—seemed to work so well. Black was contesting the widely held view that markets aspire to transmit information perfectly—for example, information about the underlying health of a company, about supply, and about demand—and that markets would thereby function better to the extent that they could instantiate noiseless information flow. Black contended, by contrast, that markets work precisely *because* of disinformation, noise, and complexity. This may seem counterintuitive, especially since we continue to read, even in our present COVID-19 moment, that the amazing performance of financial assets on Wall Street is out of sync with the economic realities of Main Street. Such reporting assumes that the price of securities and the actions of financial markets *should* be correlated with economic principles of supply and demand and that (often painful) market "corrections" are the necessary consequence of prices becoming disconnected from economic reality. Black, however, had a different vision. While he also thought that markets processed information, he asserted that they could not do so perfectly—and markets existed only because of that fact.

In making this claim, Black was reinterpreting a longer twentieth-century history of the idea that systems, and markets in particular, process information and must contend with noise. Nineteenth-century thermodynamic physics had introduced two important concepts for this tradition. The first was that some fundamental aspects of the world are stochastic (i.e., determined by chance). As a consequence, many systems change according to probabilities that can be mapped out in general terms, but the temporal trajectory of a *specific* system can never be predicted precisely. The second important principle was that every system degrades over time. This suggested to subsequent communication engineers that the amount of information in a system can be understood as the relationship of order (enthalpy) to disorder (entropy). For example, the original mathematical theory of communication invented in 1947 at Bell Labs, and upon which contemporary digital computing is based, applied ideas of entropy and enthalpy to the measure of information in a channel. Information became the probability that a piece of data would be transmitted and received at the other end of the channel. From this perspective, noise is entropic since it reduces the ability of a signal to be received.

As we noted in the introduction to this book, for earlier histories of economics concerned with labor power and energy, entropy was terminally threatening.[2] For example, for early computer scientists, cyberneticians, and game theorists who employed these theories of communication, noise was the problem, not the solution. Game theorists in the 1950s and 1960s who focused on the possibility of nuclear holocaust were haunted by the nightmare, which they instantiated in the game known as the "prisoner's dilemma," of being trapped in a room without being able to know the thoughts or decisions of prisoners in other rooms, while at the same time being dependent upon the decisions of those other prisoners. This scenario drove communications theorists, computer scientists, mathematicians, and social and behavioral scientists into a mathematical frenzy, and they sought to apply the latest advances in logic and computation to predict the future actions of the other prisoner (or prisoners) in the absence of actual information. Cyberneticians and game theorists

hoped to overcome the "dark" pools in their knowledge in order to anticipate the unknown and preempt it.[3]

In one sense, Black still employed this understanding of the relationship between information and noise. Thus, for Black, too, "noise is con-
trasted with information" and "noise is what makes our observations
imperfect."[4] Yet for Black, noise, complexity, and entropy were no longer
figures against which to battle. The overwhelming concern in the sciences of communication, command, and control in the 1950s with controlling the future by eliminating or reducing entropy had given way to a
new imagination. In this new world, chance and noise were not "devils,"
to cite cybernetician Norbert Wiener, but rather the growth medium for
markets—and if this meant that the future could no longer be predicted
or controlled, one could at least create resilient structures for absorbing
future uncertainties.[5]

One of the central technologies for capitalizing on noise in economic
markets was the Black-Scholes option pricing model, which Black developed with his colleagues Myron Scholes and Robert Merton.[6] An option is
one kind of financial derivative. Derivatives enable an investor to bet on
the future price of an underlying asset (for example, wheat, gold, a stock
price, or a house loan). The main forms of derivatives are commodity
futures, stock options, and currency swaps.[7] Derivatives are a relational
financial technology in the sense that the price to buy, and the ultimate
value of, a derivative depends on—that is, is derived from—something
else (the underlying asset). An investor who purchases an *option* has, as
the name of this tool suggests, purchased the option, but not the necessity, of buying the asset or stock at a specific date in the future for a
specific price upon which the two parties agree in the present. For an
investor, an option is attractive if they think that, by that future date, the
market asset price or stock price will be higher than the option price, as
they can then exercise the option to buy the asset or stock, immediately
sell that asset or stock at the current and higher market price, and make
a profit. For the seller, options are attractive either for the opposite reason (i.e., they think the option purchaser will have to pay more for the
stock at the option expiration date than the market price at that time)
or because the seller wants to have certainty about the price at a given
date, irrespective of market conditions (for example, wants to ensure that

when the bill for their company's debts in a foreign country are due, currency exchange rate changes will not mean paying much more for those debts than anticipated).

Though it had traditionally been difficult for traders to agree on how much the option to purchase an asset or stock should cost, up until the 1970s it was widely assumed that the value of an option would necessarily be related to the expected rate of return of the underlying stock itself, which in turn would be related to the health and profitability of the company that issued the stock.[8] Investors assumed, in other words, that an underlying objective reality—the economic health of the company— might in practice be difficult to figure out but would necessarily determine both the company stock price and options on that stock. The rational investor would thus need to gather as much information about the company as possible—for example, about management structure, supply chains, debt, and likely future prospects—and if it seemed that, objectively, the company was strong and healthy, then one could assume that its stock price would remain constant or increase. Conversely, if the company had significant problems, the market would eventually discover that truth and decrease its stock price. In both cases, options to buy the company's stock would need to be based on knowledge of the underlying economic reality of the company.

Black and his colleague Scholes introduced the Black-Scholes option pricing model in 1973 to provide a new way of answering this question of price. Contrary to the model noted above, in which options to buy stocks are linked to an "objective" truth about the company itself, Black and Scholes detached the price of an option from any expectation about the likely value of that specific underlying asset at the option maturity date. Black and Scholes instead assumed that the movement of stock prices was random, like the movement of molecules in a gas (and in fact within their investment equation used a mathematical technique for modeling the random Brownian motion of molecules).[9] Instead of basing the price of an option on the underlying objective reality of a company, the key value for Black and Scholes was the expected volatility of the stock, which meant the amplitude of the movement of the stock price up and down over time. Since volatility refers to a greater up-and-down movement of stock price than the average random motion of stocks, the volatility of one stock could only be assessed in relation to the average movement of all stocks in the market. Estimating the volatility of a stock at the time when the stock option would come due—that is, in the future—still required gathering significant amounts of information. However, the information was no longer about the specific company that issued the stock but rather about the noisy dynamics of the market as a whole. As Black noted in 1975,

My initial estimates of volatility are based on 10 years of daily data on stock prices and dividends, with more weight on more recent data. Each month, I update the estimates. Roughly speaking, last month's estimate gets four-fifths weight, and the most recent month's actual volatility gets one-fifth weight. I also make some use of the changes in volatility on stocks generally, of the direction in which the stock price has been moving, and of the "market's estimates" of volatility, as suggested by the level of option prices for the stock.[10]

The Black-Scholes option pricing model, in short, grounded the value (i.e., price) of the option in the relationship of one kind of noise (the amplitude of random changes of the stock in question over time) to another kind of noise (the random movement of stocks in general) (see figure 3.1).[11]

While Black and Scholes's quite technical and seemingly narrow approach meant that they initially encountered problems in finding a publisher for their article, once the piece was published, multiple groups began offering software for such pricing equations.[12] This was not least because Black and Scholes's premises allowed buyers and sellers of options

x – the value
In this case, the value
above the strike price

μ – mean of our normal distribution
In this case, a function of volatility, it reflects the
risk-free rate + spread of possible options over time

$$d_1 = \frac{\ln \dfrac{S_0}{K} + \left(r + \dfrac{\sigma^2}{2} \right)(T - t)}{\sigma \sqrt{T - t}}$$

σ – standard deviation
In this case, a function of volatility over time

3.1 Black-Scholes model summation. *Source:* "Black-Scholes-Merton," Brilliant.org, https://brilliant.org/wiki/black-scholes-merton/.

to come to relatively easy agreement on option prices. While it had been difficult to settle on option prices if one assumed that sellers and buyers of options had to agree on the objective health of the company, the only unknown variable of the Black-Scholes option pricing model was the future volatility of the stock, and Black's volatility charts allowed sellers and buyers to find common ground here. The success of Black and Scholes's equation was also in large part a consequence of the fact that the model joined communications and information theories with calculation in a way that made their equation amenable to algorithmic enactment.

As individuals created more complex derivative instruments that tied together many types of assets and markets, computers became essential both for obtaining data about price volatility and for calculating option prices. An entire industry, and the financial markets of today, were born from this innovation and its new understanding of noise. And because derivatives are bets on the future value of an asset, the derivatives markets can be—and in fact are—far larger than the world's current gross domestic product (GDP). (The amount of money at play in the derivatives market currently exceeds the world's GDP by 20 times.) Since the 1970s, these markets have grown massively (e.g., 25 percent per year over the last 25 years). In this sense, the derivative pricing equation is indicative of the emergence of a broader new epistemology that transformed conceptions of agency, the agent, and decision-making in the postwar period.

If the Black-Scholes model implied a new model of how to value options on stocks, it also implied a new model of the economic agent. Black and Scholes had begun working together in the late 1960s while consulting for investment firms. Their work involved applying computers to modern portfolio theory and automating arbitrage.[13] Scholes and Black began "The Pricing of Options and Corporate Liabilities," the article in which they introduced their option pricing equation, with a challenge: "If options are correctly priced in the market, it should not be possible to make sure profits by creating portfolios of long and short positions." They reasoned that since people do make money, options therefore cannot be "correctly" priced, at least in the traditional sense that the option price is a function of an underlying objective value of the stock. Mispricing—that is, the imperfect transmission of information—must be essential to the operation of markets. Noise is the only source of arbitrage.

"Rational" investing in options thus did *not* mean attempting to determine the true value of the underlying asset. Because the market is full of noise—the market exists, in fact, only because of noise—the economic agents within it cannot know the relationship between the price of an option and the "real" value of the underlying asset. However, if agents recognize these limits of their knowledge, they can focus on what they can know—namely, how a single stock price varies over time and how that variation relates to the price variations of other stocks. *Economic agency* thus meant developing mathematical tools that took the randomness, or noise, of the market as a whole into account and generated value by relating different aspects of market noise to one another.

BEYOND FINANCE: POPULATIONS AND DERIVATION IN PLATFORMS, BRAIN MINING, AND POPULATION BIOBANKS

In our third section, we will document the specific genealogy of assumptions, techniques, and technologies that enabled Black and Scholes to approach markets, options, and value in this way. However, since their option pricing equation may appear rather distant from the smart cities, phones, and other systems that are the subject of this book, we first outline through a series of brief examples the extent to which Black and Scholes's approach to derivatives and markets is structurally identical to the derivative logic of many other contemporary systems in which smartness is more clearly at play. In each of our examples below, value is generated through three key operations also central to the Black-Scholes approach to options pricing: (1) a potential source of value is located in activities in a large population that were otherwise understood as simply noise or waste, (2) a technological platform capable of linking and surveying the noisy or wasteful activities of the entire population is established, and (3) a new source of value is established by mining what was earlier understood as noise or waste. In each example, there is an effort to optimize a specific realm of activity. However, because noise is understood as ineradicable, optimization is understood as always provisional and hence interminable.

PLATFORMS

The innovation of the Black-Scholes option pricing equation was to seek value not in the valuable asset itself but rather in the noise of the market as a whole. In the same way, smartness aims to derive value from noise or waste in existing systems—or, more precisely, to create new systems on the basis of the noise or waste in existing systems. Nick Srincek and Shoshanna Zuboff have noted ways in which big data and artificial intelligence platforms often bring together two existing systems so that a third party can take advantage of waste in each of those existing infrastructures—that is, translate waste into surplus capacity, or value— and do so by linking members of a population to one another in new ways.[14] The ride-sharing company Uber, for example, was built on the model that although many people own cars that can hold multiple passengers, most people either do not drive those cars most of the time or drive them without any other passengers. At the same time, other people need to get to roughly the same location as the driver. Traditional licensed taxi services provided one solution to this problem by establishing a specific subcategory of cars and drivers who transport passengers. From the perspective of a licensed taxi system, the fact that a population of non-taxi cars and non-taxi drivers also existed was irrelevant; that is, this constituted noise from the perspective of the taxi system.

Uber's innovation was to treat that noise as a potential source of value and to see the population of non-taxi cars and non-taxi drivers as in fact wasted seats and drivers. In order to translate that waste into value, Uber had to create a platform to reconfigure the population of non-taxi cars and non-taxi drivers as quasi-taxi drivers and quasi-taxi passengers. Uber does not make the cars or the roads upon which the cars drive or even directly hire the drivers. Rather, it provides an interface (a platform) that transforms unused private cars and seats into supply, which can then meet (or, arguably, create) a previously unmet demand for auto transportation. Uber does this in the name of optimizing transportation. However, this form of optimization differs significantly from the kind of optimization that would occur in a closed system, such as that of licensed taxis, since the number of drivers and cars within the Uber system varies constantly.[15]

Lodging-sharing company Airbnb functions in the same way. For most of the twentieth century, residential homes and rental properties were understood to be completely separate kinds of accommodations: individuals lived in residential housing and vacationed or worked briefly in vacation or corporate housing. From the perspective of the vacation and corporate rental market, the fact that a residence was architecturally and functionally similar to a rental property was irrelevant. The innovation of Airbnb was to treat this noise as a form of waste that could be turned into value. The Airbnb model assumes that residences should be understood as potential vacation or corporate work properties that are not being used to full capacity (for example, when someone leaves a residence to go on vacation or a work trip). The Airbnb platform derives value from this waste by creating a population-level system of lodging seekers and lodging renters.[16]

MINING MENTAL NOISE

This search for value-creating platforms also occurs at the subjective and psychological scale. As media scholars such as Melissa Gregg and Johannes Bruder have noted, periods of unconsciousness, such as sleep, as well as moments of seeming nonproductivity are now understood as sites of potential value. The ability to approach mental states and activities in this way is in part a function of the tendency of neuroscientists to understand brains as composed of populations of networked neurons. These neuronal populations are presumed to have evolved to enable humans (and other animals) to engage aspects of their environments—for example, to learn to recognize threats and food by first distinguishing very primary forms of differences among elements of the environment (for example, edges and trajectories of objects). These basic neuronal structures evolved over the course of several million years of human prehistory and presumably have not changed at all over the last several thousand years. However, some neuroscientists have proposed that skills that are both recent and clearly cultural in nature—for example, reading and writing—in effect capitalize on aspects of the existing neuronal infrastructure in order to develop a new skill.

College de France neurobiologist Stanislas Dehaene, for example, proposes that the capacity to read letters "recycles" a more primary capacity of the human brain to recognize objects. Dehaene suggests that all writing scripts employ shapes, such as corners and vertices, that "constitute useful invariants for recognizing objects" and that when "we learn to read, part of this neuronal hierarchy converts to the new task of recognizing letters and words."[17] Dehaene describes this as "neuronal recycling" and notes that he draws in part on the French sense of *se recycler* as "students or employees who take a refresher course or train for a new job better adapted to the job market."[18] The recycling of neurons that enabled reading and human culture more generally thus turns out, in Dehaene's account, to be an example of the way in which waste within an existing population of neurons could become the basis for a new capacity. That is, if a human neuronal population no longer needed to look for threats and food every moment, it could use that same mental architecture, in combination with a system of written symbols, to enable another skill also based on identifying objects (namely, reading and writing).

Reading and writing emerged long before neuroscientific knowledge of brain architecture. The hope of many neuroscientists, however, is that knowledge of this brain architecture will enable the discovery of new forms of mental waste and, as a consequence, the extraction of new forms of value. Bruder notes, for example, that mindfulness exercises, self-monitoring and self-improvement, meditation, and daydreaming have been recast in the cognitive sciences and neurosciences as periods during which brains can optimize their mental populations to become more efficient and productive. Even dreams, Bruder notes, are understood as sites of value by the creativity industries because they allow the worker to recuperate and work better. Science ceases to view these times and states as wasteful and renders them productive by developing systems that relate noisy off time to the time of work. Employers are urged to allow workers to take time off, to enter wandering states, and to encourage meditation and other practices. A plethora of apps, which claim to manage everything from sleep to meditation, have emerged.[19] These apps seek to optimize—that is, extract algorithmically—value from what would earlier have been understood as noise, depleted resources, or waste among neuronal populations.

POPULATION-LEVEL BIOBANKS

As we noted in chapter 1, the goal of so-called individualized medicine—namely, orienting therapies and drugs toward the unique physiology of an individual—is only possible, paradoxically, on the basis of a huge volume of population-level biological samples and data. Creating a comprehensive set of deep medical files on hundreds of thousands (or better, millions) of individuals allows researchers to locate both common and uncommon relationships among a wide variety of biological differences (e.g., genetic differences) and intentionally chosen activities (e.g., smoking, exercising) and then to slot a patient into one of those individualized profiles. The key difficulty of this approach, not surprisingly, is developing the huge database of medical samples and information necessary for this work.[20]

In this case, too, success has depended upon revaluing samples and patient information previously understood as waste. In some cases this has meant relabeling an existing sample collection. The Austrian national health-care system, for example, retained biopsy samples from patients so that those samples could later be consulted if an individual patient's condition changed. The fact that the biospecimen preservation system that held patient X's biopsy samples also held millions of other patient samples was simply noise, at least from the perspective of an individual patient (i.e., the only samples that were relevant for a patient and their doctor were the patient's own samples). However, in the early 2000s researchers realized that if they linked the individual biopsy samples and associated medical records to one another, what had been noise from the point of view of an individual patient would become a source of value from the point of view of a population.

In other cases, this kind of revaluation of waste has occurred around clinical procedures. For example, researchers at the Vanderbilt University Medical Center in Nashville, Tennessee, realized that blood samples collected during regular care in patient clinics could be revalued as a site of potential biobank specimen collection because only some of the blood drawn for a clinical test was actually used and the rest was discarded. If the portion usually discarded could instead be included within, or analyzed for, a biobank, and the patient records associated with each biosample updated every time the patient visited a medical center clinic, an enormous biobank would be created fairly quickly. In this particular case, the decision was made to optimize the system and resources by retaining only genetic information from the sample. However, in other cases the entire tissue specimen itself is retained since one never knows which part of the sample—genetic information? proteomics? and so on—may be relevant and hence a source of value.

Each of these examples—the Black-Scholes option pricing model, the Airbnb and Uber platforms, cognitive mining, and population-level biobanks and individualized medicine—derive from completely different discourses and interests. Yet all four share a common logic. In each, a new system includes and revalues what was apparently noise or waste from the perspective of an earlier system and connects an entire population of members within this new system in order to enable the mining of that noise or waste. In the case of Uber and Airbnb, this means connecting a traditional rental car or vacation property model, respectively, to the much larger populations of car and housing accommodation owners or renters; in the case of cognitive mining, it means capturing and managing the noise of off-work neuronal population time to increase work productivity; in the case of biobanks, it means revaluing waste specimens and medical records and linking these to one another so that researchers can look for regularities within an entire population of samples and patients. And in the case of the Black-Scholes option pricing model, it meant treating the up-and-down movements of one stock, in relationship to all other stocks, not as irrelevant noise but rather as noise that enables two parties to agree on a price for an option on a stock.

GENEALOGY OF BLACK-SCHOLES

Each of these examples makes its specific mode of extraction, or mining, reliant on derivation in the sense that the source of value is dependent on an otherwise independent asset. In the case of Uber and Airbnb, rental value can be extracted from underlying car or accommodation ownership; in the case of brain-state mining, one manages brain states that (presumably) occur for independent biological reasons; in the case of the population biobanks examples above, value is extracted from samples obtained for independent clinical purposes; and the Black-Scholes option pricing model creates an option value that derives, but is separate, from the underlying stock price. While the Black-Scholes option pricing model is in a sense simply one among these many contemporary structures that link extraction and derivation, the fact that the option pricing model emerged earlier than these other examples, and instantiates derivation in an especially literal financial sense, allows us to use this model to clarify the assumptions about agency and temporality at work in all of these examples.

The Black-Scholes option pricing model and Black's emphasis on noise were one response to economic changes that grew out of the geopolitics of the late 1960s and early 1970s. For almost 25 years after the end of World War II, the United States enjoyed uninhibited expansion and economic growth. A series of postwar agreements, including the Bretton Woods international currency agreement (which stabilized international currency exchange rates), the Marshall Plan (which provided huge loans to European countries to assist with postwar rebuilding in exchange for increased dependence on a US-led global order), and the opening of the oil fields in the Middle East, supported an unprecedented increase in the American standard of living, the rise of a carbon-based, consumer-based economy, and a demand for US dollars and US products. Yet this system began to show signs of strain in the mid-1960s as the Bretton Woods system of international currency exchanges began to fray, and economists debated how best to resolve currency exchange volatility problems. The international currency exchange problem came to a flash point in the early 1970s with the official end of the Bretton Woods agreement. At roughly the same time, the 1973 Yom Kippur War between Israel and its Arab neighbors, and the subsequent announcement by the Organization of Petroleum Exporting Countries (OPEC) of an oil embargo on nations supporting Israel's position, further exacerbated global financial volatility. OPEC also cut oil production and raised oil prices, which led to a huge surplus in US petrodollars and even more volatility in currency and commodity markets.[21] Corporations with complex logistical enterprises were especially vulnerable to disruptions and devaluations of currencies and supply chains. This era of reduced or even degrowth was accompanied by the publication of *The Limits to Growth* (discussed in chapters 1 and 2), which increased awareness of impending environmental disaster and the uncertainty that economic growth could continue.[22]

In this context, it is perhaps not surprising that an epistemology of uncertainty emerged in finance (and also within ecology, as we document in chapter 4). Within economics and finance, this epistemology of uncertainty contested an earlier consensus among economists about the need for certainty and what was thought to be its corollary, economic stability. Directly after World War II most economists—especially those influenced by John Keynes—thought it vital for economies to remain

stable: the good economy was one without volatility, or at any rate one
that returned to its baseline growth curve. Government intervention was
focused on avoiding the sort of economic volatility that had shaped the
1920s and 1930s and had led to the rise of extremist politics in Europe.
Neoliberals, liberals, and even Socialists and Marxists had all inherited
from early eras the fantasy that chance could be tamed by statistical reasoning and science. Yet by the late 1960s and early 1970s, this no longer
seemed like an achievable goal to economists such as Black and Scholes.[23]
Counter to the geospatial claims made by the territorial understanding
of capital and eighteenth- and nineteenth-century colonial and imperial
economies, economists such as Black and Scholes no longer believed that
it was possible—or, more to the point, useful—to know fully the space of
the economy. The market not only contained ineradicable noise but was
itself only possible because of noise.

If markets were irredeemably noisy, then knowledge of markets had to
aim not at mastery, but rather at hedging risks (for example, risks of large
changes in currency and commodity prices). This intuition was shared
even by neoliberal economists who believed that, in principle, the market could be both known and stabilized. The Chicago neoliberal economist Milton Friedman had always asserted that markets could in principle
share information perfectly. However, in 1971, as we also discuss in chapter 4, he argued in "The Need for Futures Markets in Currencies" that the
collapse of the Bretton Woods currency exchange system had "creat[ed] a
major need for a broad, widely based, active, and resilient futures market"
in currencies.[24] Friedman argued that such a future market would not
destabilize the global economic order but rather ensure its stability and
resilience. But where Friedman simply proclaimed the need for such a
market, Black and Scholes actually produced new derivative technologies
predicated on volatility, noise, and uncertainty. They also recognized that
such techniques cannot control the future but must instead aim at the
more modest goal of rendering ultimately unknowable markets resilient
to shock.

Yet even as the demand for automated arbitrage and derivative pricing
models may have emerged indirectly from geopolitical realities, the inspiration for Black-Scholes's technology came from computers and the social
sciences. Their world of subjective knowledge, market uncertainty, and

imperfect information refracted a broader reconceptualization of intelligence that took place in the artificial intelligence and machine-learning sciences. It is not a coincidence that Fischer Black's degree was in artificial intelligence, rather than economics. His concepts of agents and decision-making were influenced by Herbert Simon and Allen Newell, both artificial intelligence pioneers with whom Black had initially hoped to work.[25] Black's attitude to decision-making expressed what many of these earlier pioneers in computing had begun arguing in the 1950s—namely, that our markets and machines would work better if we ceased to think of ourselves either as omniscient rational decision-makers or as Freudian beasts of irrational desires and instead embraced our limited abilities to comprehend the world by focusing on actions, not thoughts.[26] The model implemented in the Black and Scholes option pricing equation reveals a powerful distrust of human decision-making and even a distrust of human consciousness. Their computer model to automate arbitrage assumes a certain representational failure in the economic logics of both free-market rationalism and Keynesianism.

To illuminate how this understanding of decision-making and change both draws upon, but also departs from, its game-theoretical predecessors, we focus on how "agency" became "the agent" in economics. Black and Scholes's equation was derived from cybernetic and game theoretical notions of rationality. Working at MIT, Black, Scholes, and Merton all had significant exposure to the emerging cybernetic discourses and computer sciences. Black in particular was interested in figures such as cybernetician and MIT-based mathematician Norbert Wiener.[27] Wiener had worked on Brownian motion and had introduced concepts of thermodynamics and probability into the concept of communication. He is most renowned for popularizing the term "cybernetics," understood as the sciences of communication and control. Fischer and Scholes were both deeply invested in utilizing the ideas of physics and communication inherited from these sciences.

Their concept of decision-making emerged from these sciences, especially game theory, which sought to provide a means for predicting the future even in the absence of specific knowledge of an adversary's decisions. At the heart of their reformulation of how futures can be predicted was the economic theory first advanced in *Theory of Games and Economic*

Behavior, published in 1944 by John von Neumann, a Princeton mathematician and head of the Manhattan Project's mathematics division, and
economist Oskar Morgenstern.[28] As Daston and her coauthors note, *Theory
of Games and Economic Behavior* embodied and forwarded a new form of
human intelligence, which they call "Cold War rationality." They define
this form of rationality as "formal, and therefore largely independent of
personality or context," which meant that it was committed to creating
steps or algorithmic processes for decision-making and aspired to render
a decision-making process rule bound and mechanically automatable.[29]

The postwar social sciences were repositories of these techniques,
which transformed what had once been questions of political economy,
value production, and the organization of human desire and social relations into questions of algorithms and communication. Postwar neoliberal
economists, for example, began to approach markets not primarily as
sites of production, distribution, and consumption but rather as mechanisms by means of which information is managed: for these theorists, the
most fundamental operation of the market is the coordination of data,
and the economic agent is not a reasonable decision-maker but a rational
processor of information.[30] Politics, for its part, was not something separate from market information coordination but simply another mode of
the same dynamic. As Kenneth Arrow—Nobel laureate, Stanford economist, and leading proponent of rational-choice decision-making theory—
wrote in *Social Choice and Individual Values* (1951),

In a capitalist democracy there are essentially two methods by which social
choices can be made: voting, typically used to make "political" decisions and
the market mechanism, typically used to make "economic" decisions. In the
following discussion of consistency of various value judgements as to the mode
of social choice, the distinction between voting and the market mechanism will
be disregarded.[31]

Though he was writing in support of democracy, Arrow nevertheless
signaled the compression of economics and politics by way of a discourse of "choice" and "information." These algorithmic logics thus
also meant a transformation of the way in which politics was understood and an erasure of political economy in the social sciences in favor
of an interest in concepts and methods that could be used across multiple disciplines.[32]

Importantly, though, the calculative agent of game theory *was* still capable of expertise, which game theorists understood as equivalent to the capacity for optimization.[33] The key concern for cyberneticians and game theorists was the moment of decision-making, such as the moment of pressing the nuclear button (or, to take a more mundane example, choosing to buy an option). Game theorists focused on decision-making as a process and event that could be perfected. This presumed that "problems"— that is, something about which a decision needed to be made—could still be measured and bounded in time and space; end points could be measured; and one could calculate with precision the minimax solution to the problem.[34] Theorists such as von Neumann and Morgenstern never fully expelled the reasonable subject, and in fact maintained the model of an agonistic, singular, and locatable agent.

Black and Scholes, by contrast, eliminated any residue of the reasonable agent, and did so on the basis of one of the competitors of the fantasy of the technical decision-maker that had emerged in the 1940s and 1950s—namely, the idea of a networked form of intelligence. Where game theory focused on immediate decisions and was oriented toward the rational, data-driven decisiveness of the noncontemplative technocrat, Black and Scholes's alternative conception of intelligence, which set the groundwork for contemporary smartness, took some of its methods from game theory but suggested a different type of rationality, one invested in populations of agents working parallel to one another. If orthodox game theory was the way in which the field of economics "became a cyborg science," to use Philip Mirowski's felicitous phrase, this alternative approach to game theory and intelligence was how economic practices, such as the Black-Scholes option pricing equation, became cyborg technologies. In the emergent fields of finance, artificial intelligence, and business administration, the concept of rationality began to take on a new organizational form, and it was this latter lineage that proved so important for Black and Scholes.

In chapter 1 we described the version of this approach developed by Friedrich Hayek in the 1950s in his account of "the sensory order." Hayek posited a new type of networked intelligence that would permit a synchronization between individual nervous systems and the market system: a neurosocial model for economics. Hayek's conception of a networked intelligence market was grounded in the work of the psychologist Donald Olding Hebb, as well as Hebb's predecessors, cyberneticians Warren

McCulloch and Walter Pitts, who introduced the idea that neurons could be made analogous to logic gates and thus could be statistical calculating machines. In 1949, Hebb published *The Organization of Behavior*, which was to become one of the most famous and cited books in psychology (its impact in the field has been compared to that of Darwin's *Origin of the Species*). Hebb's research presented a new concept of the brain—the predecessor of what is now called "neuro-plasticity." His model was intended to counter the behaviorist approach to the brain, which posited direct correlations between "stimuli" and "reactions" (for example, Pavlov's famous dogs, which automatically reacted to the stimulus of food by salivating). Behavioralism implied that certain parts of the brain were wired to perform certain functions and that thought was merely a reaction to sensory data gathered by an organism's organs of perception.[35] For Hebb, these behavioralist premises were countered by the fact that when sections of the brain were removed or injured, either by accident or through an experiment, mental capacities or bodily functions that were initially disrupted would in some cases return or could be relearned. This suggested to Hebb that thought was the result not of hard-wired connections but rather of connections among neurons that were established during the course of an organism's experience—or, as his theory is often summarized, neurons that "fire together, wire together."[36]

The Hebbian brain model was in this sense a stochastic theory of memory and storage. The Hebbian brain was stochastic in the sense that the specific configuration of a neural net in an individual organism—those specific neurons that, because they fired together, wired together—could differ from individual to individual, or even in an individual across time. As a consequence, an individual with a brain injury could regain some functions associated with the injured area if another part of the brain could rewire itself to take on that function. This meant in turn that perceptions and cognitive activities were not stored as discrete "content" in some vast brain database, along the lines of the infinite archive of the Freudian unconscious; rather, the brain stored a process, or architecture of neural nets, and not specific pieces of data. A particular stimulus triggered a specific group of neurons to fire together, and this resulted in memory, action, or thought. That pattern could also "backpropagate" (i.e., fire in the absence of the external stimulus), which would result in the same perception, thought, memory, or imagination. This model explained classic cases of the recognition of a "gestalt" (complete form), in which the brain learns to complete an image or identify objects despite receiving only partial inputs or elements of that whole.

As many observers have noted, Hebb's model of learning was also a model of brain "programming." If intelligence could never abstract itself from its environment, this implied that if one could exert complete control over an environment, one could in principle program the brain. In 1950 Hebb tested this implication of his theory of the brain through sensory deprivation studies, which were sponsored by the Canadian Defense Research Board at McGill University in Montreal.[37] Alfred McCoy and Naomi Klein have documented the CIA's interest in these studies, and Klein has linked the CIA's use of Hebb's model of brain plasticity to what she calls the shock doctrine. This doctrine names the attempts of neoliberal theorists in the 1970s onward to apply Hebb's theory politically by employing social crises as "shocks" that could serve as occasions for "rewiring" social institutions along neoliberal principles. One of Klein's key examples was the 1973 military coup in Chile, for which neoliberal economists served as advisers, and which aimed to rewire both the psychology of the Chilean people and their institutions so that both would correspond better to what neoliberal theorists understood as the economic reality of human social relations.

Yet Hebb's suggestion that brains could never be divorced from their environments had two further implications, which tended to undercut the neoliberal approach to shock and reprogramming and were both arguably more important for the expansion of financial markets that began in the 1970s. First, if a brain's specific neural nets were always a function of its environment, this also meant that its access to "objective truth"—that is, truth understood as independent of context—was either always limited or in principle impossible. Second, Hebb's theory also stressed that if the brain could be "programmed," it was only because the brain was more fundamentally always engaged in *learning*. These two implications—that objective truth was impossible and that the brain was always learning—suggested that intelligence should be understood not as an ability to locate objective truths but rather as the ongoing ability to adjust the organism to an ever-changing environment.

This Hebbian understanding of the brain as stochastic, networked, and ecological informed the development of one tradition of artificial intelligence models. Both Hebb and Hayek, for example, were cited by Frank Rosenblatt in his development of the perceptron model, discussed in chapter 1. The perceptron was the first model of layered neuronal nets, and this model grounds contemporary machine-learning methods, including back propagation and deep-learning methods. Rosenblatt's model produced a concept of intelligence that inhered in populations and in probability; that is, it made learning a statistical and an environmental process that pooled decisions from populations in order to allow a system to adapt and learn. This approach mirrored Hebb's understanding of individual psychology and Hayek's conception of the parallel structures of nervous systems and markets.

These principles were extended in the artificial intelligence and finance fields in the 1950s and 1960s and articulated there as new concepts of organizational management and organizations. The work of psychologist and organizational theorist Herbert Simon exemplifies this transformation. Simon, who worked primarily at Carnegie Mellon University in the business school, sought to apply psychology, communication theories, computing, and cybernetics to the study of organizations and business. He was also a central figure in developing artificial intelligence throughout the 1950s and 1960s. A close compatriot of those developing models such as the perceptron, Simon sought to rethink the nature of decision-making and change in organizations along these same lines.

In 1955, Simon was working temporarily at the RAND Corporation. RAND was created after World War II by the US Air Force to introduce operations research and game theory into predictive modeling, policy, and strategic planning. Simon, then working on administrative behavior, concluded that organizations rarely act in ways that conform to classic ideas of rationality. The rules by which organizations made decisions were often idiosyncratic and failed to conform to minimax solutions, and organizations themselves often failed even to define shared problems or goals. Organizations, in short, were noisy, complicated, and unpredictable. Simon wrote that

Recent developments in economics, and particularly in the theory of the business firm, have raised great doubts as to whether [its] schematized model of economic man provides a suitable foundation on which to erect a theory. . . . I shall assume [therefore] that the concept of "economic man" (and I might add of his brother "administrative man") is in need of fairly drastic revision. . . . Broadly stated, the task is to replace the global rationality of economic man with a kind of rational behavior that is compatible with the access to information and the computational capacities that are actually possessed by organisms, including man, in the kinds of environments in which such organisms exist.[38]

Simon objected to the concept of "economic man" primarily because the rationality it purportedly possessed and employed assumed a separation between the organism and its environment, and thus presumed that the subject could process information about its system without the subject being part of, or inside, that system. He later explained his theory as emerging from his awareness of the "acute schizophrenia" that characterized the social sciences, for the latter approached individuals both as omniscient rational actors with full access to all the relevant information and, simultaneously, as affectively driven, ignorant beasts that were guided only by oedipal complexes and pleasure principles. Perhaps, Simon suggested, cybernetics made possible a compromise between these positions. Simon imagined a new subject that was capable of making systematic and apparently "optimized" decisions following preset rules but was no longer reasonable and rational in a nineteenth-century sense of possessing a perceptual field that was external to the environment in which the individual was located.

One consequence, though, was that this subject was incapable of objectivity. "We must be prepared to accept the possibility," Simon wrote, "that what we call 'the environment' may lie, in part within the skin of the biological organism." Organisms are bounded by their physiology, biology, processing capacities, and access to information. Yet organisms are still often rational in the sense that they are capable of making systematic, discrete decisions that are based on a logical order with set end points. Even nonreasoning subjects can act algorithmically—and often need to do so precisely because they lack an outside perspective on their situation.[39]

In coming to these conclusions, Simon essentially replicated Hebb's earlier psychological experiments on sensory deprivation at an organizational level. He recognized that intelligence is fundamentally environmental and can be impacted and managed only by understanding its systemic nature. In the 1960s, Herbert Simon elaborated on this vision of an agent through the figure of the ant. The ant, he argued, is only as intelligent as its environment, in the sense that it is coupled intimately with the exterior world, and its choices are determined as much by that outside environment as by the internal workings of its nervous system. The ant "deals with each obstacle as he comes to it; he probes for ways around or over it, without much thought for future obstacles."[40] For Simon, the payoff for revising the concept of agency in this way was to reframe cognitive intelligence as the production of situations and patterns for actions, rather than understanding intelligence in terms of paradigms of language or consciousness. Perception and cognition become simply modes of ant-like probing, and the idea of intelligence becomes the sum of small contextual decisions (rather than, for example, the comprehension of concepts). The orienting dream of this approach is a self-organizing system that emerges from small, rational (but not necessarily "reasonable") decisions made by very simple agents. Where liberal subjectivity had assumed a Cartesian separation of figure from ground—for example, neoclassical theories of price and markets presumed that an economic agent can model and know objectively the entire market—Simon called these assumptions into question.

Above all, Simon called into question both rational-choice decision theory and Milton Friedman's brand of neoliberalism by arguing explicitly that theories that propose clear solutions for delineating minimum-maximum "optimal" solutions fail to account for the complexity of the real world. As Simon noted, the "classical theory of omniscient rationality is strikingly simple and beautiful," and

> it allows us to predict (correctly or not) human behavior without stirring out of our armchairs to observe what such behavior is like. All the predictive power comes from characterizing the shape of the environment in which the behavior takes place. The environment, combined with the assumptions of perfect rationality, fully determines the behavior. Behavioral theories of rational choice—theories of bounded rationality—do not have this kind of simplicity. But, by way of compensation, their assumptions about human capabilities are far weaker than those of the classical theory. Thus, they make modest and realistic demands on the knowledge and computational abilities of the human agents, but they also fail to predict that those agents will equate costs and returns at the margin.[41]

Simon suggested that the neoclassical economist's belief that he could describe the market environment completely was false and therefore offered a false predictive capacity. By making the subject slightly psychotic and incapable of assigning itself a set coordinate in space and time, Simon sought to develop an approach that could better model larger-scale behaviors, such as that of organizations, in algorithmic and logical terms.[42]

Simon's organizational application of Hebb's theory of the brain provided the conceptual framework for the development of option markets and the Black-Scholes equations in the 1960s and 1970s.[43] Simon anticipated that even if the possibility of complete information is removed, one can still create markets, though the mathematics required for such an endeavor will not conform to the logical parameters and rigor of the previous era's calculative standards. Proof might not be definitive, and desires and intuitions might remain outside of logical calculation, but behavioral actions could still be calculated and programmed. More importantly, the disjuncture between the micro and macro levels of behavior—that is, the discrepancy between what individuals think they do and what networks actually manifest—was an opportunity to import the equations developed by physicists like Norbert Wiener into finance.[44] As the Swedish economist Axel Leijonhufvud has put it, "the Economic Man has given way to the Algorithmic Man"—even in economics.[45]

This algorithmic or networked intelligence, Simon argued, might make perfect prediction impossible for microeconomic theories and theories of the firm, but one could still account for stochastic processes. However, doing so required a shift in method toward processual and data-driven management tactics, which for Simon were linked to "design," or what he later called the "sciences of the artificial."[46] Simon suggested that design was a process, which produced not documentary truths about an existing world but rather artificial facts and desired outcomes. He emphasized, moreover, that the theories and practice of business management must understand organizations as dynamic, evolving, and unpredictable. By the 1970s, a term like "wicked problems" had come to exemplify the type of issues facing management. Wicked problems were both problems without clear solutions and for which the specific problem to be solved was difficult to define. Engineers, designers, and business managers regularly encountered wicked problems, which included issues such as the complexity of urban crisis, environmental degradation, and geopolitical conflicts.[47]

In finance, this networked agent created a new technology: computationally driven derivatives markets. Like ants, traders can learn and make locally optimized decisions that correspond to their limited perspective. And as in the case of ants, it is the milieux or ecology—in this case, the ecology of the market—that ultimately shapes how the future will develop and change. Intelligence is networked, evolutionary, and noisy.

Though our use of the terms "extraction" and "derivation" may sometimes seem figurative, the operations we describe above have also reconfigured literal mineral extraction and mining. This is evident, for example, in methods of gold mining in Quebec, Canada. As a geoengineer who worked at one of these mines confided to one of this book's authors (Halpern), "Quebec is for mining, what Switzerland is for banking . . . a free trade zone." The mine itself appears as a new form of zone and possibly vision of the planet's future: a landscape of automated infrastructures and communications systems dedicated to fueling commerce (see figure 3.2).[48]

Physically, the mine is a vast expanse of land. The pit itself is approximately 4 kilometers wide, and the entire mining field is 23 square kilometers. Yet despite its seeming industrial materialities, it is also a landscape of automation and smartness. Vast machines lumber through the space, carrying their rocks in stately, well-timed rhythms (see figure 3.3). These behemoths are perfectly syncopated with one another by means of a Caterpillar software platform that tracks their movements, records the load amounts and speeds with which they carry their loads, and times their loading, unloading, and maintenance. Each truck costs $3 million; each tire is 11 feet high and costs $42,000. The tires last for only eight hours of driving and then must be replaced. Though these machines dwarf human bodies, the fragility of these vast instruments seems prophetic, an omen of the limitations of both resources and time within which these mines operate.

The need for smart mining is a function both of the ownership conditions placed on the land and the paucity of gold to be found there. The mining corporation is only granted a temporary deed to what ultimately remains Canadian land (or, more accurately, unceded indigenous lands). The tenure of the deed is based on geological surveys that demonstrate the likelihood of valuable ore, and the tenure can be revised or extended only if the corporation can prove both the presence of additional deposits and the likelihood of additional value to the surrounding communities, environment, and economy. This particular mine, for example, is likely to be exhausted by 2026, which would mean that its entire life span was 15 years. At the same time, the mining company can expect to extract only one part of gold per million parts of rock; that is, the company digs and processes roughly one ton of rock (about 1 million grams) to locate one gram of gold. By the end of the mine's life it will have produced 580,000 ounces of gold but 700 million tons of waste rock and 23 square kilometers of tailings ponds (i.e., wastewater contaminated with chemicals used to purify the ore) that pose terminal acidic threats to the surrounding environment (see figure 3.4).[49] These limits on ownership and yield drive a constant effort to derive more value from the site. Every action is monitored, and every movement is optimized through sensors and live tracking of mine activity, 24-7: no stoppages allowed! The mine is constantly in action, moving in rhythms that Karl Marx might have labeled "metabolic."

3.2 Canadian Malartic gold mine, Malartic, Quebec. *Source:* Photo by Max Symuleski, August 4, 2017.

3.3 Caterpillar trucks, Canadian Malartic gold mine. *Source*: Photo by Garrett Lockhart, August 4, 2017.

3.4 Tailings ponds, Canadian Malartic gold mine. *Source*: Photo by Orit Halpern, August 4, 2017.

The paucity of gold here, and the subsequent need to employ high-end technologies and computing to locate value in what would earlier have been considered waste, is not anomalous but rather the norm for many contemporary mining operations. This represents a significant shift in the nature of mining. The Canadian Malartic mine is located in a vast expense of land, the Canadian Shield, that stretches across Canada up to the Arctic. Millions of years ago, a glacier swept out the top levels of the earth, leaving the precious minerals and metals exposed, ripe for the picking, and this area is thus considered one of the richest regions for ore deposits on earth. Yet though it took millions of years for this site to emerge geologically, the easy pickings were depleted in only a few decades, and this dynamic has been repeated everywhere in the world. As a consequence, "there is no more easy mining on earth," as the lead mine reclamation geologist at the site, Dr. Mostafa Benzaazoua, noted. He also suggested that even if we account for envisioned improvements in technology, all valuable metals and energy sources on earth will be depleted by 2155. Mining has therefore had to become smart through ever-more complicated relationships of chemistry and computation.[50]

Smartness is employed at every stage of mining, from discovery to extraction to refining. Discovering valuable ore deposits on a well-mined planet is difficult, and the search can no longer rely on the colonial appropriation of "undiscovered" land. Now airplanes and satellites provide initial information on rock formations, structures, and features, employing electromagnetic surveys, ground-penetrating radar surveys, and often satellite imagery to locate possible resource fields. Enormous resources are expended in improving the ability of artificial intelligence systems to recognize potentially valuable formations in both the energy and metal/mineral sectors, and much effort is expended on the simulation and modeling of geological formations. Once actual resources are discovered, extraction itself must also be optimized. Agnico Gold, the owner of the Canadian Malartic mine, has a new set of smart mining sites it is seeking to develop. At its showcase property, LaRonde Mine, also located in Northern Quebec a few hundred miles north, the mining is no longer open pit. Most veins are paralleled by kilometers of tunnels beneath the surface, as the corporation attempts to follow the veins of ore and derive as much value as possible from the site. Digging large tunnels and holes is complex. In order to accomplish this, new levels of artificial intelligence are applied to geological surveys and to simulation models of stability in the walls. The company has also added fully automated trucks and scoop trams and installed a new LTE-4G communication system in 2018, becoming the first mine in Canada to integrate full underground communications. This system permits drone and automated mining from remote locations.[51]

This logic of smart derivation from existing and limited resources continues in the refinement and actual extraction of the gold from the ore. When ore deposits were more plentiful, heat could be used to separate gold from rock. However, extracting gold from rock now requires cyanide, which binds with the ore and removes the gold. The same is true for most other metals, though different reagents are used; in our moment, all extraction has turned to chemistry in order to derive returns from impoverished deposits. As we noted above, this effort to extract more from less results in an increase in poisonous tailings ponds. Yet smart mining also "works," in the sense that even as the quality of ore deposits has decreased, Canadian mines produced an estimated 183 tons of gold in 2018, representing an 88 percent increase over production in 2009.[52]

Standing at the edge of the Canadian Malartic mine, one can envision what it might mean to inhabit another planet. The human capacity to terraform the earth at such scales is awe-inspiring, and this is underscored by the industrial and mechanical nature of the scene: no humans move in the pit, the truck drivers are invisible within their machines, and all other operations are automated. Yet the technical effort necessary to create such an architecture is not specific to mining, for this scene is replicated, in slightly altered form, in multiple other sites across the planet: in automated port facilities, procurement centers, and manufacturing factories. The use of smart technologies is also the rule in most other extraction industries, especially oil markets, as well as real-estate markets, smart cities, and the development of personalized medicine, to name just a few additional examples.[53] And in each of these cases, the search for value among waste has as its necessary corollary both financialization and financial derivatives. As value must be located in diminishing current assets, future behaviors become the currency of the present, with the result that we are turning the planet into a derivatives machine.

One of the many ironies of gold mining is that after all the work of smart discovery, extraction, and refinement noted above most of this gold will be placed back underground. Mine engineers noted to one of us (Halpern), for example, that over 90 percent of the gold mined at Canadian Malartic will end up in bank vaults.[54] The difference between the gold in the mine and the gold in the vault is that the latter can serve as a hedge bet against more volatile derivative and futures markets. Even after the demise of the gold standard in the nineteenth century and the end of the Bretton Woods international monetary system in the early 1970s (the latter of which had been backed by US federal gold reserves), gold remains the standard benchmark for security in producing value. As of 2010, gold markets were among the largest debt-hedging markets in the world (see figure 3.5). It is estimated that the derivative markets are betting on over 10 times the annual new mine supply of gold. The markets exceed the reality of production exponentially, setting prices and making bets far into the future.[55] As of 2015, gold markets were considered a key portion of the sovereign debt markets. Gold often serves as a "hedge" for a guarantor and a standard of value. Gold markets overshadowed national debt markets such as the Spanish debt market of the time and those of most other nations, and far overshadowed even equity investments and other hedge entities such as ExxonMobil and Apple.

In this sense, although the mines of Northern Quebec may appear to be hinterlands, they are in fact part of a *planetary urbanization*, to use Neil Brenner's term.[56] These mines, like their many sister installations in manufacturing, extraction, energy, and procurement, are an element of the global infrastructure of neoliberal economy, financialization, and globalization. Though located deep in a boreal forest, Canadian Malartic is hardly isolated from global exchanges of capital and information, for Toronto, which lies

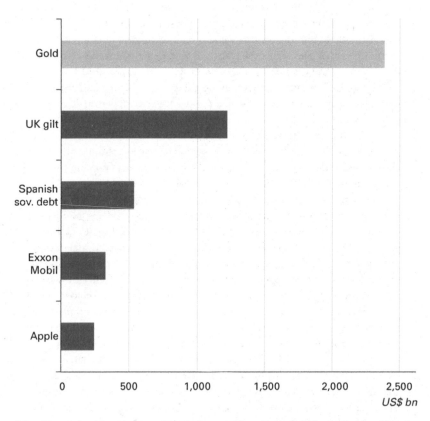

3.5 Charts of gold markets as of 2010. *Source:* Bloomberg, Bank for International Settlements, and World Gold Council, in Dickson Buchanan, "Just How Big Is the Gold Market?," August 5, 2015, https://schiffgold.com/commentaries/just-how-big-is-the-gold-market/.

some 700 kilometers to the south, is the space that coordinates the "flows" of this mine. More specifically, the Toronto Stock Exchange (TSX) is home to one of the world's largest financial markets—it is ranked the ninth-largest exchange in the world by market capitalization—and more mining, oil, and gas companies are listed on the exchange than any other.[57] Founded in 1984, the TSX has grown to $3.256 trillion of market capitalization. This financialization is the infrastructure for the folding of older extraction economies such as gold mining into the new smart and high-tech economy of the present. And, not coincidentally, Toronto is also an icon for smart city development, for it prides itself on being the symbol of the Canadian effort to turn industrial and extractive economies into artificial intelligence economies. The city regularly touts itself as a center for start-ups and innovation, a dynamic fueled by a combination of finance and large government spending on university research infrastructures, particularly in science, engineering, and medicine.[58]

THE EPISTEMOLOGY OF DERIVATION

All of the derivation techniques we have described adopt a systemic, networked—or perhaps better, *ecological*—approach to value. By this we mean that the focus is never on the underlying value of a single asset (whether that asset is a corporation, dwelling, neuron, or biospecimen, depending on the example) but rather on the relationship of the asset to a population-oriented totality (generally a market). The key question for this approach is how an asset's value changes in relationship to its risk environment. We describe this as an ecological approach to recall Simon's emphasis on the fact that decision-makers cannot be separated from their environments or ecologies. As Simon also stressed, though, as a consequence one can never be sure if any given solution to a problem—for example, a prediction of future value—is optimal.

This approach to value is, above all, an epistemological shift. We find this same epistemological shift in other discourses: it appears, for example, in the work of ecologist C. S. Holling on ecological resilience (chapter 4), and in Hayek's accounts of markets (chapter 1). The uncertainty central to this epistemology is a consequence of the noise to which Black referred and results in a concept of control that is defined as the ability to act upon, but not necessarily to know, the future. In fact, the inability to predict the future fully is what makes it possible to bet in the first place.

Black and Scholes's introduction of equations from thermodynamics into their pricing equations had the effect of naturalizing volatility and of enabling the randomness and the unpredictable nature of the market to be encoded into computational techniques. If Friedman still believed in stabilizing speculation, Black and Scholes introduced a technology that assumed that volatility is the norm and that, in order to price futures, one had to integrate models that described volatile phenomena such as the random movements of water molecules. By integrating stochastic processes with random distributions, their model in effect combined two distinct and possibly even incompatible epistemologies, one drawn from statistics (and used, among other things, to measure aspects of populations in discrete units with norms) and one from the rise of thermodynamics and indeterminacy in physics. While the other examples of this approach that we have considered here do not necessarily employ these same mathematical techniques, they are still premised on this contradictory approach to value and the future.

This combination of two histories of knowledge and of evidence and truth—on the one hand, that of the positivistic science of economics and, on the other, that of cybernetic uncertainty—has important implications for our present. These technologies naturalize and obscure social relations by making it impossible to assign causality or linear temporalities to economic events. While economists argue that risk is "out there in the world," in fact an equation such as the Black-Scholes option pricing model situates volatility within the mathematics itself, and in this way makes risk emanate from within, rather than from outside, the market. As Edward LiPuma and Benjamin Lee note, the Black-Scholes equation, by creating a situation in which cause and effect cannot be assigned, "mistakenly speaks about relational categories as though they were object categories. What this use of stochastic models thus conceals is the social processes that give rise to the phenomena." More importantly, "risks become things" that then become tradable.[59] When risk becomes autonomous, in the sense of becoming a tradable entity, this in turn displaces labor and the material world that underpins these derivatives as the site of value production. Not only are the material relations of production obscured; they are not perceived as sites of value at all, which also tends to encourage violence toward human and other forms of life. The autonomy of risk—or, more accurately, the creation of uncertainty as a commodity—can only be achieved through a new level of automation in calculation, large data sets, and computer systems capable of projecting and building more complex forms of swaps and options.

A key shift that we have stressed in our discussion of the development of artificial intelligence and machine learning is the move from *optimization*, as an act performed by fully rational or perfectly informed agents, to *derivation*, as the act of managing limited information in a broader market that one assumes cannot be fully represented or defined. Optimization requires a clearly delineated problem and parameters. Optimization and uncertainty are, as a consequence, not particularly good bedfellows (and optimization and fuzzy or ill-defined problems are even more difficult to force together). Yet one of the central features of the ideology of smartness is its emphasis on optimization to facilitate the spatial and temporal binding of problems, even as it employs derivation to facilitate the ongoing analysis and management of complex systems without temporal or spatial end points. Optimization and derivation operate as a dialectic that fuels the increasing and systematic penetration of computation into the environment. Problems are spatially contained via claims about optimization but speculatively unbound through financial instruments of derivation. In the smart city, for example, whether this is Songdo or Hudson Yards, the territory is assumed to be fully managed through big data and sensor penetration, as this will purportedly enable every city function, from electrical grids to water treatment to sewage, to be optimized in terms of cost and energy. Optimization is equally at work in remote mines, as we noted in the preceding section. The spatial organization of the mine is managed through real-time analytics and optimization algorithms. Yet the limits in materials and ecology are managed through futures options markets and derivation. Derivation thus binds algorithmically managed territories with new ideas of networked intelligence incorporated into financial technology. Smartness is the result of this combination of optimization and derivation.

Excursus 3.2
Smartness, Finance, and Racial Capitalism

Understandably, the Black-Scholes equation has been discussed primarily within the history of finance and contextualized within the emergence in the 1950s and 1960s of finance as a respectable (because finally "scientific") specialty within the field of postwar economics.[60] As our argument above suggests, such a frame is too limited for it obscures the links between this financial technology and the broader logic of derivation that we have described. It is thus tempting—but, we argue, ultimately also not correct—to address the deficits of too narrow context by resorting to the largest possible context. Cultural theorist Luke Munn's account of the Black-Scholes option pricing model exemplifies this latter approach. He (along with several other cultural theorists) understands the Black-Scholes option pricing model as a particularly "sophisticated instanc[e]" of a long-running capitalist fantasy that "if space could be more comprehensively captured and coded, it could be more intensively capitalized."[61]

Munn connects the Black-Scholes option pricing model to earlier Dutch and British colonial overseas ship-financing practices, especially eighteenth-century British slave ship insurance. For Dutch and British merchants, overseas trading expeditions—whether they concerned slaves, cash crops, or manufactured goods—were risky endeavors: ships might make navigational errors, run out of food or water, or face difficult and dangerous seas, and the slave trade included the risk that the cargo itself might revolt. The high costs of such ventures meant that money had to be raised in bonds ahead of time, and investors began merging investments in less risky and shorter voyages with longer and riskier voyages in order to hedge their bets (see figure 3.6). These charts enabled a new approach to territory, in the sense that space could be quantified and compared: the voyage from point A to point B was X times as risky as the voyage from point C to point D (with the time-based category of risk serving as the axis of comparison). As Munn has noted, this was perhaps the first truly zonal strategy for making the seas an abstract and rationalizable space.[62]

There are many differences between eighteenth-century ship insurance practices and the finance option market that interested Black and Scholes, including the basic difference between insurance for a voyage and an option on a stock. However, Munn connects these practices through their shared logic of risk hedging, which underpins his thesis that there are "racialized inequalities coded within" this "core imperative to exhaustively capitalize space."[63] Racial inequality was explicit in the slave trade, but Munn argues that it is equally present in the Black-Scholes option pricing models. The Black-Scholes option pricing models made possible the credit default swaps that underpinned the housing market bubble and bust in the early 2000s, which in turn led to a global financial implosion beginning in 2008. As Munn notes, "Hispanic

Direct Risks	1728–1729	1730–1731	1769–1770
Coastwise England	£44,510	£27,193	£65,642
Baltic	11,980	16,320	33,054
North Europe	66,085	69,486	143,988
Spain and Portugal	134,632	76,337	263,545
Mediterranean	187,910	120,715	135,589
East Indies and China	66,020	82,750	36,127
North America and West Indies	203,514	221,637	131,996
Africa and West Indies	17,250	11,732	39,688
England and Ireland	6,015	4,650	18,565
Miscellaneous	14,575	0	15,769
Total	£752,491	£630,820	£883,963

Cross Risks (not touching England)			
Coastwise Ireland	£1,024	£100	£0
Ireland and Europe and Mediterranean	77,032	26,207	14,122
Ireland–North America and West Indies	3,353	3,755	3,644
Coastwise Europe and Mediterranean	117,592	104,098	85,686
Between West Indies and North America	6,115	23,505	705
Newfoundland–South Europe	13,650	13,275	4,640
Spanish and Portuguese Colonial Trade	58,850	64,150	42,360
French Colonial Trade	4,750	5,600	0
Miscellaneous	13,545	25,001	31,405
Total	£295,911	£265,691	£182,562

3.6 A geographical distribution of risk. *Source*: Arthur H. John, "The London Assurance Company and the Marine Insurance Market of the 18th Century," *Economica* 25, no. 98 (1958): 126–141.

Excursus 3.2 (continued)

and African American neighborhoods were particularly targeted by predatory lenders—black homeowners were significantly more likely to receive high-rate subprime mortgages than white, even after controlling for factors like income and education. . . . Consequently, these communities were some of the hardest hit by the recession."[64] For Munn, the many differences between eighteenth-century ship insurance and the computerized automation of high-finance trading in the late twentieth and early twenty-first centuries should not obscure an abiding core commitment to using risk technologies to extract value from nonwhite populations.

Munn's approach is a significant improvement over finance-specific accounts of the significance of the Black-Scholes model. Yet linking the Black-Scholes model to Black Atlantic slavery so directly—that is, seeing the former as simply an extrapolation of the former, rather than anything truly new—risks obscuring as much as it reveals.[65] The first and most obvious point of distinction between the financial practices of Black Atlantic slavery and the new model of derivation inaugurated by the Black-Scholes option pricing model is the dependence of the latter on digital computing. This is a difference that makes a difference, for it has enabled modern derivatives to automate and integrate the circulation of capital and operate at new economies of scale. A second, and even more important, difference is that Black and Scholes's epistemology—that is, their understanding of markets, agents, and risk—diverges significantly from the assumptions at play in eighteenth-century British slave ship insurance practices. Eighteenth-century British slave ship insurance pricing and Black-Scholes option pricing were, as Munn notes, both oriented toward risk and what we now call portfolios, which offset high with low risk. However, eighteenth-century British slave ship insurance pricing was committed to knowing the true value of underlying assets in ways that the Black-Scholes option pricing model no longer presumes is even possible in the modern market.

The nature of contemporary racial injustice cannot be fully understood if this epistemological shift is ignored. It misses, for example, a key aspect of the dynamic that led to the subprime crisis—namely, the inscription of entire national populations into financialization by means of algorithms. Missing this point also means, we suggest, misunderstanding what it might mean to critique the modern world of noisy economies. The 2007 financial crisis was triggered by the collapse of a speculative bubble on subprime housing loans: that is, housing loans for which the risk that a borrower will default is judged to be above a certain threshold. Since the 1930s, the US government had explicitly encouraged and financially supported individual home buying, but for most of the twentieth century there was no real market for subprime loans. Banks and other lenders made a binary choice about home loans: either a would-be borrower demonstrated through their assets and personal interactions with a loan agent that they were "credible" and thereby received a loan, or they did not and so did not receive a loan. As many commentators have noted, this explicit policy was often connected to racial "redlining" policies, which ensured that people of color (especially Black Americans) would not be considered trustworthy and hence would not receive housing loans (or, when they did, would remain de facto segregated from nonwhite populations).[66]

The FICO credit score was the technology that enabled these supposedly "too risky" populations to be inducted into the national housing market. In the 1960s, racial redlining policies were successfully challenged legally, though the practice continues informally into the present. Equally important, an increasing number of elements of the housing loan application were automated into computer-assisted forms of assessment beginning in the 1970s. These measures led to the introduction, in the early 1990s, of an individual's credit score (i.e., FICO score), which most loan applicants generated through the use of credit cards and other forms of credit, into the housing loan application.[67] Individual credit scores became a central element of housing loan applications, with a score of less than 660 serving as the threshold for a subprime loan. As Martha Poon notes, the use of a numeric score for establishing the basic distinction between prime and subprime loans also enabled the latter category to be divided into different risk thresholds: that is, rather than simply establishing an undifferentiated mass of "unacceptable" (i.e., subprime) borrowers, as was earlier the case, credit scores below 660 could be broken into discrete actuarial categories, with each category linked to (and hence priced for) a certain level of risk. This in turn enabled a large subprime market, with different subprime lenders focusing on different categories of subprime risk.[68]

As a consequence of the earlier racial redlining practices, many of these subprime loans went to people of color. Though this fact—as well as the terminology of "subprime" itself—is evidence for commentators such as Munn of the continuity of this new form of financing and derivation with earlier forms of racial capitalism, this interpretation risks missing sight of how, precisely, this world of risk operates. In this case, for example, the introduction of FICO scores and the resulting finely grained actuarial divisions of risk categories into housing loans was intended to *expand*—and actually had the effect of expanding—the home-loan market. This approach presumed that risk could be properly managed only when (essentially) *all* members of the national population were included within house financing, for that in turn makes it possible to move away from seeking to evaluate risk primarily in reference to the underlying asset (in this case, an individual loan recipient's ability to repay the housing loan) in favor of technologies that engage the market as a whole.

Excursus 3.2 (continued)

Recognizing this point does not turn us away from but rather allows us to sharpen our understanding of the racial injustices at work in this case, for it focuses our attention more closely on the sites of injustice. While one key problem was clearly "predatory" loan practices (i.e., loan practices that relied on the expectation that loan recipients did not understand the nature of the loans to which they agreed), these loan practices were part and parcel of a system that assumed that a finely gradated series of housing loan risk categories could overcome earlier subjective—and hence, often racist—decisions by loan agents. Was the more fundamental problem, then, the drive toward individual house ownership? If so, the aftermath of the subprime market collapse is "solving" that in an even more problematic way, as many of the investors involved in the original crisis are now buying up foreclosed properties for an ever-expanding rental market.[69] Or was the problem instead the logic of searching for value within massive, noisy markets? If so, this is a difficult target to isolate since it is by no means restricted to housing loans but also subtends most contemporary platforms (e.g., Uber and Airbnb), population biobanking, smart cities, and cognitive mining, among many other examples.

We emphasize this point because we think it is vital for understanding the relationship of racial injustice to contemporary forms of capital. Instead of assuming that a financial logic that began in the seventeenth century extends unadulterated into the present, we instead insist on assemblages, which link together initially separate histories of reason, logic, computation, and capital. These assemblages result in a new situation, which means new forms of violence and injustice, but also new capacities for justice. For us, slavery and race indeed haunt the present, to use Ian Baucom's terminology.[70] While FICO scores, for example, may have no articulation or representation for race, it is clear that if these scores are not handled very carefully, historically poor and disenfranchised groups will be disproportionately affected. This "haunting" thus demands new forms of accounting. The FICO score is no longer racist in having a disproportionate impact on people of color in any manner that can be proven in court as causal (or specific to one group since many groups of poor people have struggled with subprime mortgages); in this sense, standard civil rights protections under the Civil Rights Acts in the US will likely not be effective. Trying to understand the new mechanisms of this economy should thus encourage us to ask new questions. For example, do we need a new, broader affirmative action policy, as exemplified by calls for a living wage, or the right to an education? Do historically disenfranchised groups need measures that adjust FICO scores?

Our point is to recognize that violence is compounded when we flatten history because the violence of new techniques is precisely that they do *not* work in the same ways as older techniques (which also means that tactics or politics that have worked historically may not work under new conditions either). The politically progressive move, then, is to recognize what *is* new while at the same time comprehending that new systems are layered on top of older histories of colonialism, race, sex, and class. This in turn requires us to interrogate how these assemblages operate and when ruptures and discontinuities emerge that can be explored for progressive causes.

CONCLUSION

Derivation, understood as an operation that extends far beyond finance, underpins our contemporary lives. With that said, though, derivatives can also encourage *critical* approaches to questions of equity and our ability to confront violence. As the critical race theorist Kara Keeling notes, contemporary algorithmic derivative practices possess a new force, and one that in is some ways no longer measurable. She suggests that the contemporary derivatives economy is one of infinite calculation without termination. The racism and injustice of such an economy, we might extrapolate from her argument, emerges from the fact that, as technologies, derivatives consume differences, whether via correlations among the value of homes in different places, the comparative poverty or wealth of different populations, the differing cost of labor in different locales, or differences in the speed by which investors can buy and sell options. These differentials become the site of a new form of automated and algorithmic speculation. Betting on differences in this way has the effect of making the future homogeneous with the present, as it perpetuates contemporary class, racial, and sexual inequities. Finance and those with capital make ever more and in fact enjoy the inequities that are now the very sites of value production. "Difference" becomes "differentials" that produce values derived from the variations between lived bodies and worlds and abstractions of finance. Racism here is not only a result of history but also a consequence of the power of algorithmic finance to eliminate time. Keeling suggests that this logic of temporal erasure can be combatted with what she calls queer and Black time, exemplified in part by Afrofuturism and surrealism. Keeling suggests that these art-political movements resist temporal erasure by remaining open to the future and to uncertainty in a manner that is plural and not homogeneous.[71]

While the consumption of differences in the name of a homogeneous and violently inequitable nonfuture—a nonfuture because the trades that determine the future have already happened in the past or present—is a vital concern, in fact these instruments also facilitate complex interactions and new forms of relationality. For example, via the FICO score derivatives permitted individuals to acquire loans despite being redlined. The same real-estate credit swaps also produced new affiliations, even if negative, between many different classes of homeowners as they collectively suffered the collapse of the market. As the Occupy movement sought to demonstrate, such affiliations may be the ground of a new politics, one that addresses the question of how value is being produced, for whom, and when.

Derivatives are the exemplars of a new form of time and a new form of value production grounded not in translation in the present but in credit to the future and uncertainty. They are not historical, but they can force an encounter with history's ghosts. In our present, not only are most minerals and metals commodity markets heavily derived but so are all energy markets—energy, in fact, is the most heavily derived market on earth.[72] The fates of pipelines and extraction infrastructure and development are in this sense often decided via a complicated set of arrangements around future bets.

However, if we leave options and derivative markets and consider the broader "smart" calculus of everyday life, the entanglements of such instruments often now make financial relations visible as never before. In their artifice and noisiness, financial technologies make social realities unbelievably visible. For example, during the COVID-19 pandemic, the inequity in the distribution of deaths among different racial groups, the unimaginable financial strain put on millions of poor and middle-class Americans, and the decreased participation of women in the workforce and the unfair distribution of household labor under lockdowns *become even more visible* when these trends are compared to the performance of the stock market (especially when one recognizes that the stock market is fueled by futures speculation). One hopes that this newfound visibility of inequity will have an impact on electoral habits, and it seems to have done so in the intertwining of the summer 2020 Black Lives Matter protests and the November 2020 US presidential election.

Some activist movements are also turning to finance as the center-piece of environmental activism. Greenpeace has recently refocused its efforts on financial entities, targeting groups such as BlackRock (the larg-est investment manager in the world) and the financial systems behind oil pipelines by shifting protests to the headquarters of the financial firm, rather than focusing efforts solely at the pipeline.[73] While in retrospect this may seem like common sense, in fact, until recently the financializa-tion of the energy sector was not a focus of activism in the way that the actual corporate perpetrators of environmental damage were.

Within the industries of prediction and insurance, there have also been changes. Recently, Swiss Re, in collaboration with the Nature Conservancy, has developed a new initiative titled "underwriting nature," which creates instruments to insure coral reefs and natural "assets." The Mesoamerican Reef of the Yucatan is the test subject for this approach. The Mesoameri-can Reef is valuable for coastal defense, biodiversity, and food, and the intent is to produce an instrument that could create payments in the case of severe hurricanes. Local communities, the Nature Conservancy, private companies, and the federal government have started a trust to pay the pre-mium, which is funded via taxes from tourism and government sources, along with (potentially) investments and donations. The express hope is that insuring the reef will make the latter economically visible as an asset and therefore worthy of preservation in the minds of many.

The insurance itself is intended to provide funds to "repair" the reef when it is damaged, though occasions for insurance payouts are limited. For example, the insurance protects against hurricanes (and is automati-cally activated by the speed of the winds) but does not cover reef damage and related economic injuries in the case of water acidification due to climate change. Critics argue that the fund turns nature into a commod-ity and transfers money to a private firm (Swiss Re) rather than allowing those funds to go directly to preservation. Those in favor of this approach, though, argue that is the only way to get the money to repair the reef, and because the insurance is not paid all at once (money is collected from different groups and populations over time), the reef can be preserved in a way that was impossible to achieve earlier.[74] According to Veronica Scotti, chairman of Global Partnerships at Swiss Re, "What we see is a new type of parametric insurance product that offers rapid disbursement of capital, which can be adopted for broader application in the market. We believe this could become a very effective tool to help countries protect their oceans better and achieve climate resilience faster."[75] Resilience and finance are always bedfellows, as we will see in chapter 4, but here we must ask how the new parameters serve as both opportunities and hazards for the future.

Similar evidence can be found in bond markets, where financial instruments can serve as sources of political action. For example, developing climate preparedness mechanisms for large American cities (a topic to which we return in the next chapter) is very expensive. While finance is usually depicted in urban history as an attempt by neoliberalism to bankrupt the public, numerous recent examples suggest other possibilities.[76] In Houston's reconstruction after Hurricane Harvey, for example, a unique bill structuring the bond instruments was passed in 2018. This plan used the money from bonds to assist those poor and minority neighborhoods that were at highest risk for flooding. The bond was structured so that different forms of risk were assessed differently, rather than merely evaluating risk through property value and property loss. Evaluation through property loss costs alone would have benefited rich neighborhoods. While there are many ongoing battles, its success is not yet clear. Federal funding was slowed under President Donald Trump, and the bond structure and financing became a site of political action.[77]

Similar trends have been documented globally. While scholarship is only beginning to document these new directions, James Mizes has noted in his study of bond issuing in Africa (particularly in Gabon, Ghana, and Senegal) that there is an emerging trend toward "domestic capital markets" that take international development expertise and make it a site of African "unity, collective development, and ownership." While there is no question that such markets are about privatization, they have also leveraged a growing middle class and potentially might open routes to new forms of development that are not solely controlled by former colonial powers in the region.[78] Anthropologists such as Janet Roitman have also begun to look at infrastructure financing and remittances as new financial tools that transform debt and credit ratings for West African nations, permitting access to money in global capital markets. While the effects of these developments are still to be determined, for her it is critical to realize the place of such techniques in shaping planetary finance and the possibility of leveraging such technologies to create infrastructure and transform life (for good and bad). For Roitman, such new strategies gesture at new forms of political economy.[79] These authors all stress that neoliberalism is not homogeneous and—whether it was intended or not—sometimes produce new forms of publics and politics.

In a less corporate model, the Green New Deal proposed in the US Congress has as a central feature the creation of new financial technologies and leverage instruments that fund widespread green infrastructure projects and support the transformation from carbon to renewable energy. These instruments are to be developed and leveraged "in a way that ensures that the public receives appropriate ownership stakes and returns on investment, [and] adequate capital." In the Green New Deal, this ideal of leverage is conjoined with a desire to produce new technologies and practices that will encourage and create green (and, we presume, smart) infrastructures. What smartness denotes here remains a political question, but in these documents, technology, greenness, and equality are linked. Our challenge, then, is to explore the potential of smartness for more equitable forms of living and to diversify our understanding of what it means to be smart, whether that means including Indigenous knowledge, rethinking the ownership and distribution of new technologies, or deciding which technologies to develop and forward.[80]

The issue of how these instruments will operate returns to a fundamental question about the relationship of politics and economics first posed by Kenneth Boulding, systems theorist and perhaps the first popularizer of the idea of "Spaceship Earth." Boulding's question was: How can accountability and different forms of time and relationality be inserted into the logic of deferring risk, and therefore into the modes of responsibility inherent in the logic of derivation? In a famous article, "The Economics of the Coming Spaceship Earth," he imagined a world now forced into viewing itself as a closed system. This meant understanding the world as ruled by the second law of thermodynamics, which in turn meant facing increased volatility and resource depletion and accepting the inability to access any external resources through which one could evolve and change. Under such conditions, Boulding suggested, we must find ways to transform measure, and space, into time. Within an economy that is open in time but closed within the territory of the planet, new concepts of accumulation—of knowledge, well-being, and diversity—must be created.[81]

Cultural theorist Randy Martin's approach to derivatives might be one way to create these new understandings. Martin has urged us to see derivatives not as instruments separated from the social processes of production and reproduction but rather as tools that demonstrate an increased interrelatedness, globalization, and socialization of debt.[82] That is, by openly revealing the fabrication and relationality of value and exposing the operation of derivation (whether in financial markets or in social networks or smart infrastructures), these instruments make the interdependence and sociality of value visible and therefore a site of intervention and action. Derivatives make us more indebted to each other—and also to the earth, which is often the subject of such exchanges, whether via weather future markets or real-estate options—by tying together disparate actions and objects within a single assembled bundle of reallocated risks to trade. The political and ethical question, then, is: How might we activate this increased indebtedness in ways less amenable to the strict market logics of neoliberal economics? Our response to these questions, as we have noted earlier, relies on genealogy for the creation of histories that recognize that hauntings are one way of disrupting the seamless assumption that the future will look like the present. Perhaps this is one important lesson of smartness: that uncertainty is not something to be defeated through certainty but rather something to be embraced.

4

RESILIENCE

In the midst of the COVID-19 pandemic, invocations of both a present and future "new normal" circulate ad nauseam throughout news outlets and social networks. This new normal has multiple, and often contested, dimensions, denoting the likelihood that social distancing protocols will remain in place for many years to come; that app-based health monitoring and access will become even more central to daily life; and that increasing precarity for many and a dramatic increase in profits for a small few will continue unabated, to name just a few proposed aspects of the new normal. Ubiquitous curve graphs and data visualizations help us to grasp these dimensions of the new normal.

This language invokes not only a now vanished "old" normal, which becomes an object of nostalgia, but also encourages us to reconsider the concept of the normal itself. As Georges Canguilhem, Michel Foucault, and François Ewald, among others, have noted, the concepts of norms, normativity, and normalization came to prominence in the eighteenth and nineteenth centuries and were bound up with, and to, concepts of economy, population, and race. Invocations of the normal in these discourses were also necessarily claims about nature, even if, as Foucault stressed (following Canguilhem), the nature referenced by practices of what he called "normativity"—that is, the use of the human sciences to locate new possible norms and "nudge" social relations toward those—was understood as pliable and capable of change. Though we may now believe that nature and culture cannot be rigorously distinguished and that we live in a modulatory, postnormal, postanthropocentric, and posthuman society, the invocation of the new normal emphasizes the continuing importance of this ideal of a nature that would enable both the old and the new normal.

But if the language of the new normal contains an implicit reference to nature, what form of nature is this, precisely? We can begin to approach this question by considering calls for cities or states to "flatten the curve" (see figure 4.1). The goal and language of infection curve flattening emerged from a 2007 US Centers for Disease Control and Prevention article on the community-based mitigation of pandemic influenzas.[1] This discourse has

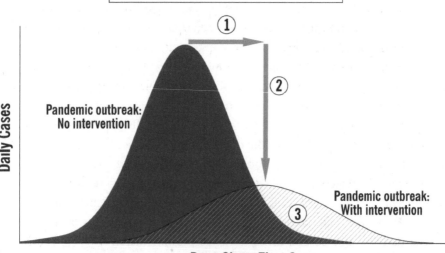

Goals of Community Mitigation

① Delay outbreak peak
② Decompress peak burden on hospitals / infrastructure
③ Diminish overall cases and health impacts

4.1   Chart illustrating CDC plans for managing a pandemic, published more than a decade before the start of the COVID-19 pandemic. *Source*: Centers for Disease Control and Prevention, *Interim Pre-pandemic Planning Guidance: Community Strategy for Pandemic Influenza Mitigation in the United States—Early, Targeted, Layered Use of Nonpharmaceutical Interventions* (Atlanta: Centers for Disease Control and Prevention, 2007), 18.

some curious features. The assumption that we *will* end up on the curve and that the best we can do is to flatten it assumes that pandemics are inevitable, although it is uncertain when and where they will start. Such a premise was prepared both by public health officials' warnings about coming pandemics, by the emergence of infectious diseases such as HIV/AIDS in the Global North, and by popular nonfiction books and movies with titles such as *The Hot Zone, The Coming Plague, Outbreak,* and *Contagion* (not to mention the slew of zombie-apocalypse films and television shows that fill our imaginaries with the logic of infection). Virtually no one in public health doubted the possibility of another zoonotically transferred pandemic; the only question was when. This means, in turn, that though pandemics are clouded in uncertainty—*Which* disease will it be? *When* will it hit? *From where* will it emerge, and *to where* will it spread?—they are still understood as events that will inevitably need to be managed. Pandemics are, in short, "known unknowns."

The discourse of curve flattening also assumes that efforts to mitigate the consequences of new infectious diseases will always be significantly hampered by human-created obstacles that seem in principle preventable. Public health professionals and many others fully understand that better urban planning, more social equity, stronger public health infrastructures, transformed agricultural systems, and improved environmental management would likely prevent many future pandemics. Yet few actually believe that these infrastructural changes will occur, no matter how many lives such measures would save. As a consequence, the best we can do is to manage this uncertain event (i.e., flatten the curve). Because COVID-19 spreads through the act of breathing, we have to slow the metabolism of the social system so that we can accelerate the demise of the virus. This is an example of the management of temporalities: a strategy that assumes catastrophe *will* occur but that there are ways to address this trauma. Those ways, as mentioned in chapter 2, largely involve an army of tracking apps, the construction of ever more digital surveillance, and, of course, new modes of working, learning, and living online.

Our primary goal in this chapter is to trace the genealogy of (1) the premise that environmentally induced trauma should not be understood as either an event or shock, but simply itself the new normal, and (2) that, as a consequence, humans must develop "resilient," data-intensive structures that can transform what earlier would have been considered a catastrophe into a managerial possibility. Our approach borrows from the work of Lisa Parks and Janet Walker, who discuss *disaster media*. For them, disaster media is a heuristic that allows us to understand how environmental and "natural" catastrophes are coproduced with media infrastructures, which in turn creates new forms of governmentality, narrative, and inequity.[2] We trace here the emergence of the contemporary concept and practices of resilience, with resilience understood as that which enables management in the face of constant change and crisis; demands large-scale and distributed information gathering; emphasizes speculative scenario planning (which itself requires financial speculation); and, as a result, conceptualizes the planet and life itself as test beds for demo-ing possible futures.

Our account has three parts. In the first we document how the development of ecosystem ecology between the 1940s and 1960s helped to reconfigure approaches to the environment in ways that would subsequently make it graspable as a medium for computation and speculation. Ecosystem ecology itself, though, did not take this step, and in the second section we focus on the emergence of the notion of ecological resilience in the 1970s, stressing that this concept explicitly critiqued the concept of homeostasis central to ecosystem ecology, and implicitly challenged the idea of "limits" promulgated in the report *The Limits to Growth*. In the third section, we use the example of *adaptive management* techniques to trace the merging of ecological concepts of resilience with business practices, with the treatment of ecologies as "service providers" acting as a key link between these discourses. Recalling our examples from chapter 2, our conclusion to this chapter notes that smartness is often oriented toward the goal of making urban spaces and other infrastructures "resilient" to climate change in a manner that makes some populations vulnerable and expendable while maintaining the wealth and power of others (a dynamic exemplified in figure 4.2). This chapter not only aims to understand how this itself became the new normal but also seeks to point out alternatives to this vision.

FROM SYSTEMS ECOLOGY TO ALLOPLASTICITY

On July 16, 1945, in the New Mexico desert, the first nuclear device was detonated (figure 4.3). The result of one of the most massive scientific

Dow mounts a comeback

Dow Jones industrial average from February 12 to March 2

SOURCE: FactSet. Data as of market close on 3/2/2020.

4:01 pm: Dow surged nearly 1,300 points

4.2 Dow rally on Monday, March 2, 2020. *Source*: CNBC, "Dow Rallies Nearly 1,300," accessed June 2, 2020, https://www.cnbc.com/2020/03/02/stock-market-today-live.html.

4.3 Trinity test fireball at 16 milliseconds after detonation, Alamogordo Bombing and Gunnery Range, New Mexico, July 16, 1945. *Source*: Wikipedia, s.v. "Trinity," last modified February 6, 2022, https://de.wikipedia.org/wiki/Trinity-Test#/media/File:Trinity_Test _Fireball_16ms.jpg.

efforts on earth, the Trinity test would enable the design of the bomb named Fat Man, which was dropped on Nagasaki, Japan, three weeks later. The test, in other words, almost immediately ceased to be a test and became a reality.

After witnessing the Trinity test explosion, J. Robert Oppenheimer, the scientific director of the Manhattan Project, quoted the *Bhagavad Gita*: "If the radiance of a thousand suns were to burst into the sky, that would be like the splendor of the Mighty One." As the large mushroom cloud bloomed over the desert, another line from the same scripture came to his lips: "I am become Death, the destroyer of worlds." Oppenheimer would soon turn publicly against his own invention, unable to stomach the fact that he had helped to construct a technology that would shatter the world—that is, a machine designed for nothing but death and that, simply through its testing, had already transformed all life on earth.[3]

The Trinity test marked a pivotal moment when species survival and technology were intimately, and horrifically, intertwined. Radioactive fallout from the test ensured that every element of life was penetrated by the effects of human technology, which is also to say design. The very shell of the planet was transformed geologically by radiation. Today, the radioactive trace of this test is one of the main markers used by the geological sciences to demarcate the newly proposed geological era of the Anthropocene. This test defined the moment when human materials and technologies entered the earth's crust and could be scientifically measured. It also heralded the start of a new technical era that has reshaped the planet's climate and geology. And by helping to end World War II with a US/Allied victory, the Trinity test inaugurated the rise of American empire and the start of the Great Acceleration and the Information Age. All of these developments were driven by the new energy and computational machines unlocked through the war. In this sense the Trinity test marked the moment when technologies and life could no longer be separated and when design could be understood as techne for transforming human life at a planetary scale—sometimes through computation and calculation; sometimes by using populations, both human and animal, as media; and even more often by linking computation and populations.

Oppenheimer's reflections on the outcome of what remains to date one of the largest and most technically intensive and expensive design projects in human history would be uncannily repeated by the designer Victor Papanek some 30 years later to describe a new feature defining the human. Humans, Papanek argued, cannot be distinguished from nonhuman animals by their possession of either language or toolmaking. The nascent and interrelated sciences of ecology, sociobiology, cybernetics, and ethology had discovered, for example, that bees have languages, and other animals construct vast architectures. Rather, Papanek argued that

mankind is unique among animals in its relationship to the environment. All other animals adapt *autoplastically* to a changing environment (by growing thicker hair . . . etc.). . . . Only mankind transforms earth itself to suit its needs and wants *alloplastically*. This job of form-giving and reshaping has become the designer's responsibility.[4]

Humans, Papanek argued, make climates rather than adapt to environments. Humans can indeed destroy worlds, as Oppenheimer noted, but for Papanek, this possibility was grounded in the more fundamental human capacity to *make* worlds.

Papanek's claims were part of a more general rethinking of the concept of the environment in the 1960s and 1970s. Where environment had earlier tended to be understood as either a set of forces external to the body or as the habitat within which living beings found themselves, Papanek conceptualized environment as a medium akin to other media, such as film, photography, or metal, that could be given form and reshaped. Papanek was among the first to conceptualize what we will call the *planetary test bed*. We adapt this term from Papanek's colleague and inspiration, the architect Richard Neutra, who referred to the "planetary test," though we modify the phrase to underscore the technical and engineering aspirations of this approach.[5] Within this vision, the planet and all its populations—of humans, information, materials, and nonhuman lives—are reenvisioned as a milieu for the growth of computation. This also means rethinking human life and habitat as an experiment and opportunity for design intervention and growth at a planetary scale. Design, Papanek implied, should no longer focus on adaptation to an environment understood to be outside human control but rather should be understood in terms of a more fundamental capacity for alloplasticity—namely, the forming and reshaping of the earth. It is our gift, or curse, to deny adaptation and instead desire technical and design interventions.

Our goal in this section is to document the steps—primarily, though not exclusively, in ecosystem ecology—that led from Oppenheimer's horror at the irreversible effects of the Trinity test on the planet itself to Papanek's enthusiasm about the possibilities this enabled for rethinking our planetary condition. Our first stop on this path is a key linkage established in the late 1940s among nuclear testing, risk management, and the study of ecology.

Between 1948 and 1958, the United States used the string of newly acquired Marshall Islands to conduct roughly 50 nuclear weapons tests. These so-called Pacific Proving Grounds ended up being ground zero for both nuclear technology development and for the birth of a cybernetically informed ecology. These tests unleashed massive amounts of radiation into the surrounding environment and seas, affecting the flora, fauna, and people of this territory. With the usual arrogance of empire, Americans treated the irradiated Indigenous people and animals as resources for developing new technologies and knowledge. Dr. Robert Conrad, the doctor in charge of testing and medical care for the hundreds of Marshallese on the islands who had been exposed to radiation, suggested in a 1957 memo that the Islanders could "afford most valuable ecological radiation data on human beings."[6]

As Elizabeth DeLoughrey notes in her brilliant analysis of the Pacific Proving Grounds, this epistemology of power was underpinned by the "myth of the isolate." As DeLoughrey and Richard Grove have documented, both the concepts of Eden and "the deserted island" were critical in enabling the rise of both modern science and colonialism. Since at least Sir Thomas More's *Utopia* (1516), the isolated island was understood as a space located outside of European culture and society that had its own set of rules. In More's *Utopia*, the isolated island was a mirror world for possible political reorganizations of Europe. Through the figures of the Pacific Islands and the New World, islands served as critical imaginary spaces that allowed Europeans to imagine that they were encountering others who either were not human or at least were not as human as themselves. The island could therefore legitimately serve as a laboratory, a place where conditions not normally encountered in the mundane European world could be found or induced.

Perhaps paradoxically, this fantasy of a space from which European social order was absent and in which one could encounter "nature" sat comfortably with projects of terraforming and transforming these supposedly isolated, pure, or primitive habitats into more civilized ones. From the draining of swamps in Palestine to combat malaria, to the transformation of agricultural systems in India, to the dredging of ports and rivers to expand trade, empire, terraforming, and geoengineering have long been accomplices of one another. Terraforming is not a contemporary discovery. The civilizing mission included taming nature and making it economically productive in the terms set up by plantation systems, mineral and energy extractive economies, and, later, industrial agriculture.[7]

Beginning in the late nineteenth century, ecology had a particularly important place within these sciences of empire. "Isolates," whether they were literal islands or geographies that could be treated like isolated islands, were central to the development of ecology: Charles Elton's seminal concept of the food chain, for example, was based on his research on Bear Island, a small arctic island that contained fewer than 100 animal species, while G. Evelyn Hutchison (whom we discuss at more length below) developed many key ecosystem notions through his study of Linsley Pond in New Haven, Connecticut.[8] The premise of the ecological isolate, as historian of ecology Joel B. Hagen notes, was that the ecological processes located in the isolate "were comparable to those operating in the biosphere as a whole."[9]

Early twentieth-century ecologists also often adopted a mechanistic approach to nature and these isolates. Historian of science Peder Anker argues that the British naturalist Arthur Tansley, who coined the term "ecosystem" in the 1920s, "believe[d] . . . that a complex system like the human mind or society could be explained in terms of simple biological processes, which in turn are based on physical and chemical laws of energy."[10] Tansley also understood nature as something that could be managed in an industrial and colonial manner. The study of ecosystems became integral to the mechanical maintenance of the machineries of empire. Colonial administrators who took inspiration from the new concepts of ecosystems came to naturalize the order of the British Empire as the reflection of the order of nature. Concepts of evolution, apex ecosystems, and holism, all of which were central to late nineteenth- and early twentieth-century ecology, reinforced the idea of Britain's ascendance; in this sense, models of nature justified models of governance.[11]

In the post–World War II period, nuclear technologies enabled the concepts of the island isolate and the ecosystem to combine in a new way, for exposing a purportedly isolated island to radioactivity enabled the development of a cybernetic imaginary of the ecosystem. More specifically, nuclear tests provided ecologists with a new way to see the relations between living beings by providing scientists with a new form of inscription.

Howard and Eugene Odum, among the most successful ecologists of the era, visited the Marshall Islands from 1954 to 1955 in order to study the Eniwetok Atoll, part of the Pacific Proving Grounds. Howard Odum had focused in his dissertation on the movement of strontium in the environment, while Eugene had investigated the ecosystems around the Savannah River Site in Aiken, South Carolina, where nuclear materials were processed for weapons. The coral reefs of the atoll presented the ecologist brothers with an imagined isolate and extreme laboratory. They understood these reefs as extremely varied ecosystems that, "save for [minor] fluctuations," seem "unchanged year after year, and reefs apparently persist . . . for millions of years." Their primary question, then, was the following: "How are steady state equilibria such as the reef ecosystem self adjusted?" Yet immediately after noting their interest in equilibria, they turned to a massive change: "Since nuclear explosion tests are being conducted in the vicinity of these inherently stable reef communities, a unique opportunity is provided for critical assays of the effects of radiation due to fission products on *whole populations and entire ecological systems in the field.*"[12] The bomb produced a new expanse of testing and a new test bed at the scale of the ecosystem, but it also opened the fear of change. Instability and disruption could now happen at speeds not imagined by ecologists, who envisioned systems stable for eons, during which evolution progressed at steady but slow speeds. The Odums' work would pioneer a new form of ecological perception that combined materialities and methods bridging the emerging sciences of communication and control: cybernetics, systems theory, and computation.[13]

For the Odums and the ecologists who followed their lead, radiation inscribed the movements of materials and energy throughout ecologies in ways that enabled these movements to be visualized and quantified. Radiation was soon to be equated with energy and metabolism and hence became a method for visualizing food chains in a system. In making these metabolic relations visible, research at these grounds opened the way to thinking of life at a planetary scale in terms of information or data transfer (understood as quantities of energy) and also made these complex networks both visible and capable of being modeled. In this sense, the ecological work done throughout this period laid the foundations for the subsequent computational modeling and designing of environments.

Howard Odum was particularly important to this process, for he extended significantly the role of concepts and practices of feedback, cybernetics, and simulation into ecosystem study. Odum was a student of Yale ecologist G. Evelyn Hutchinson, who attended the Macy Conferences on Cybernetics in New York in 1946 and was among the first to begin applying systems theories emerging from the new sciences of communication and control to ecology.[14] In his paper for the conference "Circular Causal Systems in Ecology," Hutchinson introduced what he described as biogeochemical and biodemographic approaches to ecology. The biogeochemical approach sought to merge the study of biology with that of geology. Hutchinson suggested that this would be a largely quantitative approach, which would employ flowchart-like diagrams to document the relations of energy and chemistry among different parts of a system (see figure 4.4). The biodemographic approach was more conceptual and did not measure anything, though it was grounded in abstract mathematical models that purported to describe population growth and behavior; these models were intended to reveal the conditions "under which groups of organisms exist . . . [and were] self-correcting within limits."[15] As the historian of science Peter J. Taylor argues, "For Hutchinson whether ecology was biogeochemical or biodemographic—it was nevertheless united by a theoretical proposition: Groups of organisms are systems having feedback loops that ensure self-regulation and persistence."[16] For Hutchinson's cybernetically informed version of ecology, systems were composed of feedback loops that facilitated adaptation, survival, and stability.

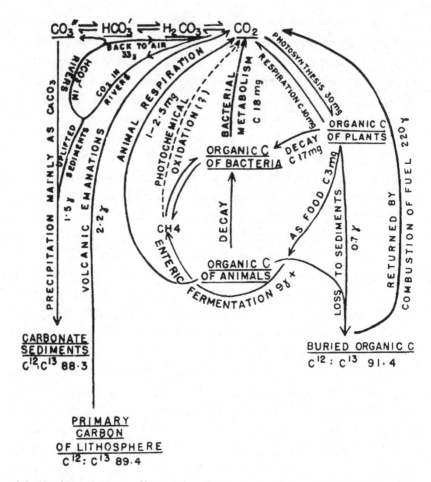

4.4 Hutchinson's image of biogeochemical processes. *Source:* G. E. Hutchinson, "Circular Causal Systems in Ecology," *Annals of the New York Academy of Sciences 40* (1948): 221–246.

Howard Odum extended these concepts of his mentor throughout his work. His research on the Eniwetok Atoll, for example, established a paradigm of mapping the relations between organisms in ecosystems and producing representations and models of the system. Radiation played a key part in making visible the metabolic cycles—for example, phosphorus, nitrogen, and carbon cycles—and energy consumption of systems. Odum went on to conduct a vast body of research, including a major study in which parts of a tropical forest in Puerto Rico were irradiated in order to examine the impact that radiation and technology have on ecosystems. As Odum argued in 1983, his overarching research goal had been

to develop a systems language that would combine features of actual systems, drawing from other systems languages as needed. An energy circuit language of symbols and diagrams was developed combining kinetics, energetics, and economics. It does mathematics symbolically but at the same time keeps track of energy laws. In the process of its use it was realized that the diagrams are themselves a form of mathematics with emergent theorems and perceptions for the workings of the mind that extend the capacity to see wholes and parts simultaneously.[17]

Odum was in this sense a pioneer of a new form of technical vision that permitted local measurement to take on planetary or scalar proportions and thus bridge the very small and the very large—or, as he put it, the "wholes and parts simultaneously."[18]

Part of Odum's technical vision was to model parts of the world through generative flowcharts that trafficked in the same language of electrical engineering that underpinned early computing and computer programming (figure 4.5). He imagined and built analog "machines," so to speak, that mapped energy by means of a visualization practice analogous to programming and organizational charts of the time and were visually like circuit boards. Odum took ideas of information and combined them with energy and feedback to begin modeling how systems self-regulated. These diagrams became both simulations of future ecosystem behavior and epistemologies of organization.

The concept of energy was central to Odum's project of approaching ecosystems through the lens of engineering concepts. Borrowing from practices of circuit design in electrical engineering, Odum treated energy as a common unit for representing processes between discrete elements or units within ecosystems. Energy was bound to ideas of feedback in engineering, and representing ecosystems in terms of energy enabled one not only to represent these systems but to do so in a symbolic language commensurate with, and translatable into, digital computation. One cost of this approach, however, was that the separate ecosystem elements that the ecologist identified—for example, a specific animal population—were treated as homogeneous units. In treating ecosystem populations in this

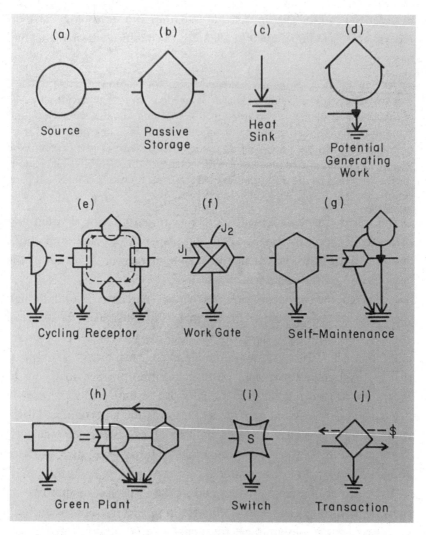

4.5 Energy-circuit "language" for modeling ecosystems. *Source*: H. T. Odum, "The Rain Forest and Man: An Introduction," in *A Tropical Rain Forest*, ed. H. T. Odum and R. F. Pigeon (Oak Ridge, TN: U.S. Atomic Energy Commission, Division of Technical Information, 1970), A-6.

way, Odum explicitly drew on the mathematician and statistician Alfred J. Lotka, who himself had drawn on thermodynamic physics to develop a series of equations in which the members of populations were treated like identical particles of gases.[19] Although this approach allowed Odum to treat the populations of an ecosystem as nodes within a circuit or network, it also moved interest away from those questions of change, novelty, and speciation that, as we noted in chapter 1, were central to the population thinking for which Ernst Mayr advocated.[20]

One consequence was that Odum's flowcharts allowed ecologists to envision the world as calculable, computational, and dynamic only because they also presumed that ecosystems naturally tended toward homeostasis. The concept of *apex ecologies*, for example, was a way of acknowledging change but containing it within the frame of homeostasis. One ecology would give way to another, but this would eventually lead to—or at least tend toward—a final stable "apex" ecology. For Odum, the coral reefs of the atoll were such a system, for they could remain "unchanged year after year . . . for millions of years." More generally, we can say that the premise of "isolating" ecosystems for study, and imagining that a system could be modeled and its future managed in its entirety, tended to position *change* as a disturbance, rather than an endemic condition. This did not mean ignoring change: Odum's work on the coral reefs, for example, was made possible by the human introduction of new radioactive isotopes into the environment. But it did mean understanding ecology as the study of the means by which ecosystems resolve change into stability.

This simultaneous interest in change and the valorization of homeostasis and stability was not specific to Odum but rather integral to the cybernetics discourse upon which he drew. For cyberneticians and the many disciples of midcentury operations research and communication sciences, the world *was* fundamentally stochastic, prone to accidents and unforeseen events, and thermodynamic entropy would always prevail in the long run. At the same time, cybernetics presumed that feedback and careful engineering would enable pockets of self-organization and regulation that ran counter to the inevitable degradation to which the second law of thermodynamics pointed. The key, then, was to keep change and disturbances small and continuous, and so within the operating limits of well-engineered systems. The real problem, from this perspective, was volatile and dramatic change and disruption, which was incommensurable with homeostasis. In the realm of political analysis and policy, for example, cybernetics-inspired analysts and policy-makers felt confident that game-theory models could keep in check the "closed" competition between the world's two nuclear superpowers, the US and the USSR. However, they feared that the decentered and networked world that was emerging in the wake of decolonization and economic transformation in the 1960s would not allow for this kind of stability and homeostasis.[21]

THE LIMITS OF LIMITS: FROM THE CLUB OF ROME TO RESILIENCE

Howard Odum's valorization of homeostasis within ecosystem ecology had implications leading in two quite different directions. On the one hand, Odum's understanding of ecosystems as made up of discrete components that could be represented in flowchart fashion and as tending toward stability and homeostasis enabled him to develop a sense of both the connections among, and the vulnerabilities of, the parts of the planetary ecosystem. He was a pioneer in recognizing that energy and its metabolism had implications for both humans and all animal life and in encouraging the belief that this planetary metabolism could be modeled and perhaps even optimized by means of machines. This vision—though often developed by Odum with the help of military funding—provided tools for identifying environmental injustices; the threats that industry posed to various ecosystems; and the costs of human violence, such as war, that resulted from the military-industrial complex (see figure 4.6).[22] Equating social and environmental engineering by means of both the model of the flowchart and concepts of connectivity and communication thus meant an ability to envision *changing* social and natural environments, both for the better and for the worse. In producing a shared language between computation and environment, Odum's innovations paved the way toward making ecology a medium for design.

On the other hand, Odum's cybernetic-derived emphasis on stability and homeostasis meant that intentional changes to the world's social and natural economics could only be understood as virtuous if such changes were oriented toward a stable, "final" state that, like the coral reefs, could persist unchanged. The possibility of *constant* change was,

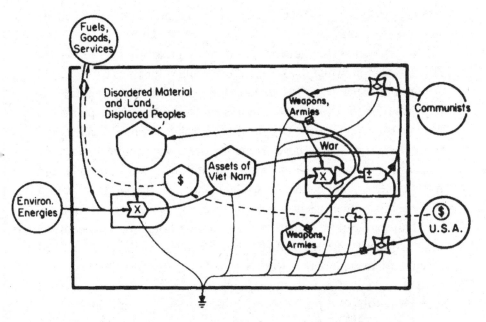

4.6 Howard Odum's model of the Vietnam War. *Source*: Howard T. Odum, *Systems Ecology: An Introduction*, ed. Robert L. Metcalf and Werner Stumm (New York: John Wiley & Sons, 1983), 552.

from this perspective, not thinkable as anything other than a slide toward entropy. Yet it was not clear how such a perspective could take into account the view of some ecologists that many ecosystems did *not* appear to stabilize after suffering disruption. Simply ending the use of a toxin or reseeding an environment, for example, often did not return the system to its past state, and even seemingly environmentally friendly actions, such as lowering fishing quotas or replanting trees, had little effect once certain levels of disruption to the ecosystem were surpassed.[23] Even more problematic, it was not clear how the homeostasis view could deal with the fact that massive disruptions to regular ecosystem behavior—whether from human technologies such as DDT or nuclear blasts, or in the form of past "natural" mass extinction events—were, though extreme, nonetheless relatively standard natural events, at least when one considered long enough timescales. These facts suggested that systems might be *naturally* volatile. Though Odum's ecosystem ecology could only view these facts through a negative lens, some ecologists wondered whether change could be understood as a potential, a source of value, and a necessary activity in nature.

The first of these implications of Odum's approach was evident in *The Limits to Growth*, while the second set of implications was developed in critiques of that report.[24] As we noted in chapter 1, the computational dimension of *The Limits to Growth* was led by Jay Forrester, a pioneer in systems thinking and the design of large-scale computer systems. Forrester worked at MIT to develop the first large-scale computer systems, such as the Semi-Automatic Ground Environment for antiaircraft and nuclear defense, and he had subsequently applied computing to a range of social and human science problems, such as managing industrial supply chains and urban design. While returning home in a plane after a meeting in 1970 with the Club of Rome group in Bern, Forrester sketched a flowchart diagram of the "world system" since the year 1900 and afterward received funding to create a simulation that would model the world system's possible futures.[25] Closely replicating the type of diagrams produced at the time by figures such as Odum, Forrester drew the world in terms of flowcharts and computer programs.[26]

Donella H. Meadows, the lead author of *The Limits to Growth*, remembered Forrester's contribution as stressing that the key problem facing the world was "*growth*—exponential growth of energy use, material flows, and population against earth's physical limits. That which all the world sees as the solution to its problems is in fact a cause of those problems."[27] At a 1968 meeting of the Organisation for Economic Co-operation and Development in Bellagio, which included many future members of the Club of Rome (and also Forrester), Hasan Ozbekhan, director of planning at the System Development Corporation, suggested that humankind was facing "Continuous Critical Problems," which included pollution, poverty, and racial discrimination.[28] What made these problems difficult to address was the human inability to think systemically and globally. Problems seemed to have no limits, but human thinking did. *The Limits to Growth* sought to solve some of these problems via computing, suggesting that while humans simply could not comprehend scale or nonlinear growth, machines and their models were able to do so (figure 4.7).

As we noted in chapter 1, *The Limits to Growth* sold millions of copies, and its success underscored the extent to which a global public was open

Figure 24 FEEDBACK LOOPS OF POPULATION, CAPITAL, AGRICULTURE, AND POLLUTION

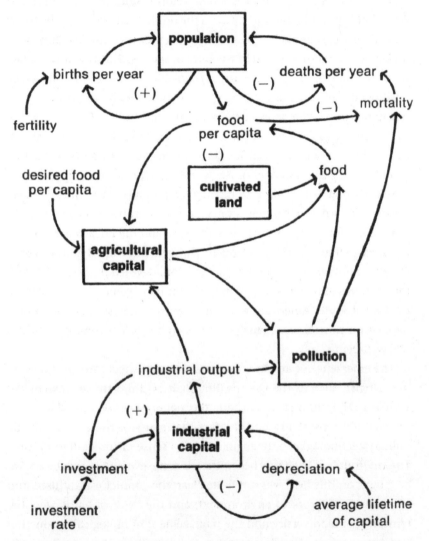

4.7 Imagining world problems through feedback loops. *Source*: Donella H. Meadows et al., *The Limits to Growth: A Report for the Club of Rome's Project on the Predicament of Mankind* (New York: Universe Books, 1972), 97.

to its message about system limits, the dangers of surpassing those limits, and the protocols of computational modeling of the world system. The report relied on a widespread mood, developed since the end of the Second World War, that the world was on the brink of catastrophe. The threat was initially that of nuclear weapons, which produced the first large-scale imaginary of a terminal end of the human species. This threat was channeled into a contest of the warring ideologies of the Cold War, which positioned the planet as a closed space, within which only communism or capitalism could emerge as the sole winner.[29] In the 1960s, texts such as Paul Ehrlich's *The Population Bomb* added a second fear—namely, that the newly decolonized Global South and its imagined masses of population would deplete the resources of the world. Postcolonialism suggested the possibility of a world split into fragmented and warring entities, some of which were armed with new strategies of both subversion and population. Major foundations, including the Ford and the Rockefeller Foundations, as well as the US government through the United States Agency for International Development organization, devoted millions of dollars for programs aimed at studying population management and control in the Global South. Americans especially came to understand international population control and management as crucial to ensuring their health and well-being.[30]

The emergence of an environmental consciousness through the work of ecologists such as the Odums illuminated a third threat. Insecticides such as DDT, fertilizers, and other industrial products seemed able to disrupt natural ecosystems, and the fact that many of these were initially military technologies only augmented the sense of impending disaster. The environment itself had become a war zone, even if the "adversary" to be fought on this field was not quite clear; the conflict was diffused into a worldwide network of environment; and the fields of battle were the territories of biodiversity and the management of biological production and reproduction. Drawing together these multiple threats, *The Limits to Growth* amplified an already existent European and American sense of threat that bound together population, environmental, and political issues and that, it seemed, could only be addressed via the instantiation of some sort of homeostatic "steady state" for the world system.

Yet this emphasis of the Club of Rome and ecologists such as Howard Odum on homeostasis also provided impetus to a quite different emergent discourse that we label *resilient hopefulness*. While this discourse was as technocratically optimistic as the Club of Rome and Odum, it understood "limits" quite differently. Where Odum and the Club of Rome employed computation in order to produce purportedly objective knowledge of those limits beyond which systems could not function, the discourse of resilient hopefulness employed a new epistemology, one that emphasized that uncertainty could never be eliminated. It thus also pointed to a new technology of *management* that aimed not at control but rather at resilience in the face of endemic shocks. We can observe the emergence of the discourse of resilient hopefulness in two fields that had an especially strong stake in the uptake of *The Limits to Growth*: economics and ecology.

HAYEK, FRIEDMAN, AND RESILIENT MARKETS

Since the stability of the post–World War II international system of Western-style democracies had been premised on the constant economic growth of national economies, economists were likely to be concerned by a report that advocated, on ecological and demographic grounds, for an end to growth.[31] For many economists, this implied that one would have to choose between ecological stability or political stability; one could not have both. In his 1974 speech for his Nobel Memorial Prize in Economic Science award, the economist Friedrich Hayek disparaged *The Limits to Growth* as part of a general plea, addressed to both mainstream economists and their leftist critics, for a more modest epistemology that would give up on the dream of complete knowledge of and control over the future. Hayek noted drily that the recent creation of the Nobel Prize in Economic Science was itself testimony to the "propensity [of economists] to imitate as closely as possible the procedures of the brilliantly successful physical sciences" but stressed that, in economics, this often "led to outright error." Hayek stressed that economies were *not* equivalent to the isolated systems of physics. This was in part because a social science such as economics focused on the behavior of large populations of different agents, with the result that like much of biology but unlike most fields of the physical sciences, [economics has] to deal with structures of *essential* complexity, i.e. with structures whose characteristic properties can be exhibited only by models made up of relatively large numbers of variables. Competition, for instance, is a process which will produce certain results only if it proceeds among a fairly large number of acting persons.[32]

Rather than pretending to be able to replicate the kinds of discoveries about the natural world available to physicists, economists should instead accept a biology-like world of uncertainty, chance, and large populations of different individuals. This would mean relinquishing the goal of *planning* and turning instead to the more modest goal of *managing*. For Hayek, societies emerge from decentralized networks of information coordinated through markets, which meant that seeking to plan or regulate the economy—by, for example, limiting or eliminating growth—could only end in disaster.

Hayek suggested that mainstream economists, by seeking to emulate the physical sciences, had in fact given encouragement to precisely that fantasy of control he saw as central to *The Limits to Growth*. He suggested that

it is often difficult enough for the expert, and certainly in many instances impossible for the layman, to distinguish between legitimate and illegitimate claims advanced in the name of science. The enormous publicity recently given by the media to a report pronouncing in the name of science on *The Limits to Growth*, and the silence of the same media about the devastating criticism this report has received from the competent experts, must make one feel somewhat apprehensive about the use to which the prestige of science can be put. But it is by no means only in the field of economics that far-reaching claims are made on behalf of a more scientific direction of all human activities and the desirability of replacing spontaneous processes by "conscious human control."[33]

For Hayek, systems self-organized from the "free efforts of millions of individuals" and not the conscious decision-making power of the few. As a consequence, control—understood as predicated on the prediction of future events, whether by mainstream economists or the Club of Rome— was impossible. For Hayek, though, this was not cause for despair. Rather, it was grounds for hope, provided that those populations of millions were allowed to engage new and unanticipated problems flexibly by means of unrestricted market activity.

Hayek's lecture focused primarily on the rather abstract realm of epistemology and provided relatively little guidance as to what this approach might look like in practice. However, in the 1970s several economists and ecologists turned to concepts of flexibility and resilience to explain how the epistemological modesty valorized by Hayek could generate solutions to specific new and unanticipated problems while at the same time avoiding system collapse. Within international relations, one such problem was the failure of the Bretton Woods international currency exchange system in the late 1960s and early 1970s. The Bretton Woods system was designed shortly after World War II and was supposed to keep Western economies stable by preventing large international currency exchange rate fluctuations, which many economists and policy analysts saw as one of the key causes of the rise of Fascist and totalitarian regimes after World War I. However, the system—which pegged international currency rates to the US dollar and the US dollar to a fixed gold exchange rate—was having serious problems in the 1960s and finally ended in 1971, when President Richard Nixon declared that the US dollar could no longer be exchanged for gold.

Chicago School neoliberal economist Milton Friedman saw in the collapse of the Bretton Woods system an opportunity for creating a new, resilient system of international currency exchange. In a 1971 article titled "The Need for Futures Markets in Currencies," Friedman acknowledged that, in the absence of an international system of currency controls, exchange rates would shift constantly in relationship to one another. The architects of Bretton Woods had seen such volatility as a problem since it meant that those engaged in foreign trade would have to take significant risks that the currency in which a trade was negotiated would depreciate by the time payments were to be made. Bretton Woods thus sought to institute a "system of rigidly fixed [exchange] rates that do not change." However, as Friedman noted, they ended up with a "system of rigidly fixed rates subject to large jumps from time to time," and these large jumps eventually broke what was designed to be a rigid system of control.[34] Friedman argued that the solution could not be another rigid centrally controlled system but instead a resilient futures market for currencies: that is, there was a "major need for a broad, widely based, active, and resilient futures market" that would allow those engaged in foreign trade to hedge the risks associated with currency exchange changes.[35]

For Friedman, "resilience" was to be understood as the opposite of "rigidity," and would mean, in practice, something like the oxymoronic notion of "stable change." More specifically, currency markets would change in response to global events but nevertheless continue to protect international trade, the international global political order of the West, and the primacy of the United States within that order. Although Friedman was presumably one of those economists chastised by Hayek in his lecture as overly committed to "scientific" models of economics, Friedman's proposal for resilient futures markets nevertheless exemplified Hayek's image of markets that flexibly managed, rather than rigidly controlled or planned, an always uncertain future.

HOLLING, ECOLOGY, AND RESILIENCE

The terminology of resilience was also at the center of a new discourse in ecology, one that subtly contested both Howard Odum's commitment to homeostasis and the implementation of that vision in *The Limits to Growth*.[36] Ecologist C. S. Holling began his 1973 essay "Resilience and Stability of Ecological Systems" with a contrast between two ways of looking at the world:

INDIVIDUALS DIE, POPULATIONS DISAPPEAR, and species become extinct. That is one view of the world. But another view of the world concentrates not so much on presence or absence as upon the numbers of organisms and the degree of constancy of their numbers. These are two very different ways of viewing the behavior of systems and the usefulness of the view depends very much on the properties of the system concerned.[37]

Odum and the Club of Rome valorized a world without change (i.e., stability) and so understood change only as either the movement toward stability or the first step toward collapse and catastrophe. Holling, by contrast, sketched a view of the world in which change—even catastrophic change—is the norm. Yet, Holling proposed, such change leads not to the end of systems but rather to their evolution. Changes may indeed cause some species to go extinct, yet systems themselves, "degrees of constancy," and evolution persist. Holling used the term "resilience" to capture this latter capacity of systems (figure 4.8).

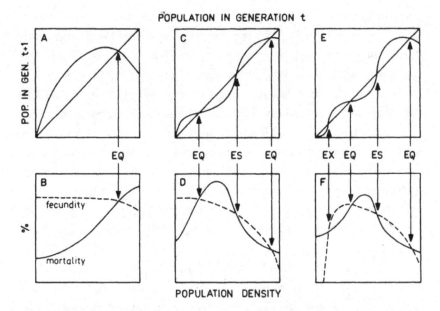

POPULATION IN GENERATION t

4.8 Diagram speculating on various futures for population reproduction curves and deriving fecundity and morbidity (*bottom row*) from these curves. *Source*: C. S. Holling, "Resilience and Stability of Ecological Systems," *Annual Review of Ecological Systems* 4 (1973): 1–23, 21.

Holling based his new term, "resilience," on a number of ecological facts that Odum's systems approach had difficulty incorporating. In an early critique of then-current models of industrial fishery and forestry management, Holling demonstrated that using insecticides, reseeding lakes with fish, or replanting one type of tree did not in fact return ecological systems to an earlier, purportedly stable state. Holling's research on budworm infestations that destroyed economically valuable softwood forest stands (conducted while he was employed by the Canada Department of Forestry in Sault Ste. Marie, Ontario) led to an even more startling conclusion.[38] The spruce budworm is a small defoliating insect that had plagued the boreal forests of North America in periodic episodes for centuries. In response, in 1951 the Canadian province of New Brunswick initiated an insecticide-spraying program. In the short term, this successfully reduced tree mortality. Yet these efforts did not appear to work in the long term. Using historical data from 1951 to the early 1970s, Holling's group discovered that forests went through cycles of fluctuating populations, changing their entire state and character over longer sweeps of history; that is, there is no one "apex ecology," for forests constantly shift among dominance of spruce, birch, and balsam fir trees (figure 4.9). Under natural conditions, budworms contribute to this process of change. It was in part the *tree industry's preference* for one type of tree over another that encouraged the notion of an "ideal" state of the forest that should remain stable.[39] Holling's computational modeling of the historical time series also suggested that tactics intended to maintain this ideal state, such as the use of insecticide, actually increased the area of the forest vulnerable to incipient outbreak conditions. In other words, attempts to maintain stability were, over the long term, counterproductive since they extended vulnerability and the possibility of future outbreaks over larger territories.

Managing ecosystems with a focus on stability was, in short, an error. Efforts to distinguish taxonomically between populations, count the size of each population, and place these numbers in flowcharts of linked boxes that retained their form because they were connected to one another by processes of negative feedback fundamentally misunderstood the nature of ecologies. Positive, rather than negative, feedback was the central concern for Holling's approach to system modeling, for positive feedback produces dynamism and change. This emphasis on change suggested to Holling that one must think of ecologies not primarily in terms of their resident populations but in terms of the *processes* or *services* that an ecology provides. What, in fact, does a forest *do*? Does it provide trees? Shade? Hatching grounds for other species? These processes were the central elements of ecologies, and these processes had to be distinguished from the numeric counting of populations, however important the latter might also be. If humans wanted to manage forests, they had to seek to maintain these processes, rather than simply a specific number of individuals in a population.

In the case of the forest in the budworm studies, for example, Holling suggested that the absolute number of spruces is not important. What *is* important is the ability of the forest to rejuvenate and to continue growing trees, and this capacity depends upon fluctuating numbers within specific tree and insect populations. Better ecological management meant understanding that systems change: forests in Ontario, for example, might be used for some time for leisure and vacationing and then for forestry, and their management must change accordingly. As we note in more detail below, Holling described this as "adaptive management" and argued that while this approach necessitated constant gathering and analysis of data, the goal of data gathering was not to locate deviations from a stable state but to enable managerial goals and methods to change in response to a constantly changing ecology.[40]

If in the 1940s and 1950s the nuclear bomb and its fallout had provided a new way to visualize ecological relations, by the late 1960s contamination of the environment made it possible for ecologists such as Holling to reenvision those relations not in terms of stability but in terms of constantly mutating systems. Resilience denoted for Holling the capacity of a system to persist by changing in periods of intense external perturbation. The concept of resilience enabled a management approach to ecosystems that "emphasize[s] the need to keep options open, the need to view events in a regional rather than a local context, and the need to emphasize heterogeneity."[41]

In order to secure more possible routes for adaptation in case of unanticipated shocks, environmental managers had to create multiple strategies for future action, think "regionally" (that is, in terms of networks and connections across different territories and times), and emphasize heterogeneity (e.g., biodiversity).[42] Resilience was defined in relationship to crisis and states of exception; as we noted in the introduction to this book, resilience can be a virtue only when crises are assumed to be either quasi-constant or the most relevant issue for managerial actions. Holling underscored that the movement from stability to resilience required an epistemological shift: "Flowing from this [emphasis on resilience] would be not the presumption of sufficient knowledge, but the recognition of our ignorance: not the assumption that future events are expected, but that they will be unexpected."[43]

Seeing the world through the lens of resilience meant not only expecting the unexpected but also employing the unexpected as occasions for *learning*. As Holling noted,

Efforts to reduce uncertainty are admirable. . . . But if not accompanied by an equal effort to design for uncertainty and to obtain benefits from the unexpected, the best of predictive methods will only lead to larger problems arising more

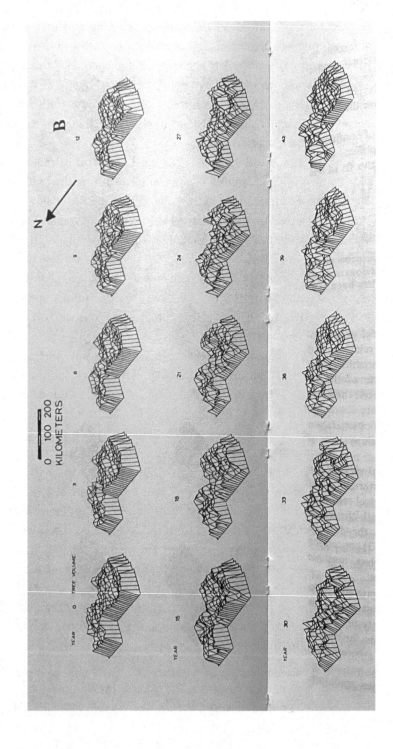

FIGURE 11.8   Spatial behavior of the budworm–forest model under historical harvest and spraying rules. The coordinates are as defined for Figure 11.7. The orientation and scale of Figure 11.8A are the same as in Figure 11.8B. Figures 11.8A and 11.8B show patterns of egg density and tree volume, respectively, beginning with conditions known to exist in 1953. Compared to Figure 11.7, the management policies can be seen to preserve trees, but at the expense of creating permanent semioutbreak conditions, highly sensitive to policy failure.

4.9   Topological models generated from historical data since 1951 of budworm population densities in space. It is also worth noting that these new dynamic maps and capacities to compare data sets came with the introduction of digital computation and new platforms such as the Canadian geographic information system (CGIS), considered the root of contemporary GIS systems in the early 1970s. *Source:* C. S. Holling, "The Spruce Budworm/Forest Management Problem," in *Adaptive Environmental Assessment and Management,* ed. C. S. Holling (New York: John Wiley & Sons, 1978), 143–183.

quickly and more often. This view is the heart of adaptive environmental management—an interactive process using techniques that not only reduce uncertainty but also benefit from it. The goal is to develop more resilient policies.[44]

Adaptive management was a method for transforming ecology into engineering while at the same time transforming engineering into an adaptive learning process—or, in Papanek's terms, channeling the human
capacity for alloplasticity into environmental management.

As our reference to Papanek underscores, Holling's theory of resilience
was in one sense simply symptomatic of larger shifts in ideas of evolution, as applied to both nature and culture, that occurred in the 1960s
and 1970s. For example, new theories of evolutionary change, such as
the Gaia theory developed by chemist James Lovelock and microbiolo-
gist Lynn Margulis in the early 1970s, focused on the capacity of living
organisms to metabolize the geological, energetic, and chemical materials
of the earth to induce climatic change. Breaking down the clear separation between the geological and the organic, Gaia theory was predicated on regular extinction events and massive pollution scenarios as one
microbe or geological event after another transformed the environment
into an inhospitable milieu for fellow organisms. Yet despite the regularity
of massive extinction events, "life" persevered and constantly evolved,
and was in this way always able to reformulate environments.[45] Holling's
emphasis on epistemological modesty was echoed in Hayek's market theory, while economists such as Eugene Fama sought to demonstrate in
the 1970s that extreme volatility and so-called fat tail distributions were
much more common in financial markets than would be predicted by
theories that presumed a steady state normal curve in market growth.[46]
Holling was, from this perspective, simply one of many who contested
the view that economy, culture, and life were characterized by stability.[47]

If we have focused on Holling's theory of resilience, it is because this
theory, especially its implementation within the practice of adaptive
management, has proven to be especially important in drawing together
the seemingly separate realms of ecology and economics. We will expand
on Holling's key notion of adaptive management in the next section, but
before doing so we want to make three summary points about resilience.

The first is that because resilience assumes uncertainty and volatility
as our common, perhaps even "normal," condition, the life and death

of specific individuals or even populations is in principle less important than the ongoing evolution of systems. Second, resilience refers to a new way to model systems and therefore to define and measure their elements. Instead of developing taxonomies that enable ecologists to organize populations into stable categories, resilience encourages ecologists to define systems in terms of processes and to measure the relationships *between* populations and other factors (nitrates, carbon, energy, and so on). A first corollary of this second point is that the techniques designed for ecological management also apply to human systems since there is no hard and fast distinction between these two kinds of systems.[48] A second corollary of this second point is that past data can be used to build concepts and encourage experimentation but can never actually predict the future. Our third point is that ecologists interested in resilience emphasize heterogeneity and diversity as important to facilitating resilience. Systems without a *surplus* of functions and populations cannot adapt. Perfectly optimized systems would collapse when faced with change—and the latter was inevitable.

There are tensions among these summary points. On the one hand, the focus on processes and what today are called *ecosystem services* necessarily means that some lives and populations can be acceptably sacrificed so long as the system continues to operate; in this sense, trauma is a regularized and normalized event. (In later ecological work on hierarchies and models of ecosystems, ecologists often prioritized "key" species or relationships in order to make the model operative.[49]) On the other hand, environmental managers recognized that only systems with robust diversity, redundancy, and supplemental capacities might survive abrupt and catastrophic events—which meant, when combined with Holling's emphasis on epistemological modesty, that it was impossible to know in advance which lives and populations were "actually" disposable. Resilience thus oscillates between the two poles of Darwinian evolutionary theory: on the one hand, survival of the fittest; on the other, the need for diversity within and between populations in order to enable adaptability. *Optimization* of the system always potentially comes at the cost of adaptation. This means, though, that ecology can contest contemporary forms of artificial intelligence grounded in the assumption that improvements always occur through optimization of the "fit" of solutions to a representable model.

ADAPTIVE MANAGEMENT

Holling's new understanding of ecological systems as made up first and foremost of services (rather than, say, populations), his emphasis on epistemological modesty, and his stress on enabling resilience for an always uncertain future brought his vision of ecology quite close to the neoliberal understanding of markets. It is thus perhaps not surprising that Holling's understanding of resilience has spread far beyond ecology proper and has become central to disaster management, design thinking, humanitarian aid, governmentality, and infrastructural security; in addition, his related concept of adaptive management has moved outward from ecology to fields such as business and urban design.[50] By 2007, Holling himself had become a central figure in both environmental policy and business management circles, not least because of his work founding the Stockholm Resilience Center, a major university and international science advisory council that encourages research on resilience and socioecological systems, works closely with numerous United Nations branches, and develops initiatives for the global seafood industry.[51]

While other scholars have noted the spread of concepts of resilience from ecology into other fields, they have generally not focused on the extent to which Holling's distinctive approach to data gathering, computer modeling, and *scenarios* have been part of that package. We want to stress this point, for it is the roles of data and model building for adaptive management that have made the latter especially important to the development of smartness and the smartness mandate. In his field-defining textbook, Holling urged ecological managers to accept that "comprehensive 'state of the system' surveys (species lists, soil conditions, and the like)" are *not* "a necessary step in environmental assessment." Holling stressed that

survey studies are often extremely expensive yet produce nothing but masses of uninterpreted and descriptive data. Also, they seldom give any clues to natural changes that may be about to occur independently of development impacts. Environmental systems are not static entities, and they cannot be understood by simply finding out what is where over a short survey period.[52]

Moreover, ecological management encompassed for Holling both social and environmental features that are difficult to define and which therefore escape the earlier and static models of ecosystem ecologists such as the Odums.[53]

In order to negotiate these limits of large data sets, two practices began to emerge within both ecology and business management. On the one hand, managers moved away from the goal of predicting the future and instead adopted methods of scenario planning for extreme events. In his foreword to Holling's *Adaptive Environmental Assessment and Management*, environmental manager Martin Holdgate echoed Hayek's critique of efforts to *predict* the future, contending that in the case of environmental studies, "much effort has been devoted to . . . collection of unnecessarily large quantities of data that have given rise to undue expectations and unsatisfactory predictions."[54] Holling and his colleagues argued that though large data sets are often necessary for effective environmental management, such data should be used to create models that are understood from the outset as partial and always in need of future refinement (or even abandonment). Models enable the development of multiple scenarios of the future, but the intuitive plausibility of these scenarios can also be used to contest aspects of the model.[55] Whether Holling and his colleagues drew the term "scenario" from Cold War planners such as Herman Kahn, their approach paralleled the development in the late 1960s and early 1970s of scenario-planning divisions within major corporations such as Royal Dutch Shell. These corporate divisions also emphasized the impossibility of predicting the future and the importance of using scenarios recursively as a means of allowing companies to learn and adapt to changing conditions.[56]

There is a significant distance between Holling's vision of fallible scenarios and the models of the future provided in *The Limits to Growth*. Meadows and her colleagues also stressed that their "graphs are *not* exact predictions of the values of the variables at any particular year in the future" and noted that the model could and should be refined as more data was gathered.[57] At the same time, Meadows and her coauthors stressed that *their model itself* was complete, in the sense that it accurately captured and revealed the world system's "behavioral tendencies":

Even in the absence of improved data, information now available is sufficient to generate valid basic behavior modes for the world system. This is true because the model's feedback loop structure is a much more important determinant of overall behavior than the exact numbers used to quantify the feedback loops. Even rather large changes in input data do not generally alter the *mode* of behavior, as we shall see in the following pages. Numerical changes may well affect the *period* of an oscillation or the *rate* of growth or the *time* of a collapse, but they will not affect the fact that the basic mode is oscillation or growth or collapse.[58]

Additional data would thus enable only a refinement, rather than a fundamental disqualification, of the report's model itself.

Adaptive management (and scenario planning) contested precisely this belief in the unquestionable validity of the model. Adaptive management thus sought to study and collect data on *management systems themselves* in order to develop more complex models of their operations so that this understanding could be fed back into, in order to improve, management principles. Resilience managers must bring together better knowledge about how systems survive stress with an awareness that elements of systems will always escape current models. As the limit points of computational analysis and prediction, scenarios not only described possible futures but also served as a constantly shifting and new frontier for calculation. The scenario was both what one could not in fact model but could only guess at intuitively, and what ongoing efforts at data gathering and feedback models must constantly seek to capture.

Principles of adaptive management have been integrated into many contemporary business practices, often by merging with practices that have different genealogies. In his work on business continuity management (BCM), for example, Andreas Folkers demonstrates how notions of ecological resilience entered business management and informed disaster and catastrophe preparedness in the case of the financial sector.[59] Folkers notes that BCM

is a nascent disaster preparedness and recovery strategy that is mostly applied in the private sector. It seeks to ensure the continuous functioning of the most fundamental business processes in the face of various emergencies. It involves establishing redundant infrastructures like relocation sites and back-up systems, as well as the preparation of emergency protocols to enable swift and effective responses to disruption. BCM is a generic risk management strategy, but it is especially prominent in the finance sector. This is because 9/11 not only heightened awareness of the importance of proper disaster response strategies in the financial sector, but also highlighted the shortcomings of existing incident management strategies.[60]

Folkers notes that though the genealogy of *continuity planning* can be traced back to Cold War efforts to ensure the persistence of government and government services in the event of nuclear war, this aspiration was then linked to concepts of resilience in the early 2000s. If adaptive management aims to enable *learning with continuity*, BCM is focused on ensuring the continuity side of that aspiration.[61]

The principles of BCM were made stunningly visible in 2012 when the Category 3 hurricane Sandy hit the New York metropolitan area. The storm devastated the infrastructure there, leaving almost 10 million people literally in the dark. Infrastructure damage was particularly intense in minority and Black neighborhoods, as well as public housing projects in zones such as Red Hook in Brooklyn.

Yet in the midst of the storm, a single gleaming building in Manhattan *did* have power, despite the outages everywhere else. That building was investment bank Goldman Sachs, one of the major financial institutions in the United States. Its ongoing operations were both in support of and an example of adaptive management tactics cultivated after September 11, 2001, to ensure the ongoing operation of its financial services. If one were to imagine the ecosystem services of lower Manhattan, it is clear that—at least in the vision of certain planners and engineers—*finance* is the key service, and much of the rest of the city can essentially be sacrificed to maintain the continuity of this function. Goldman Sachs's continuity managers are, in any case, more concerned with continuing services than anticipating the specific disaster that might strike the city. Continuity management aims to overcome logistical challenges, and over the last decade, Goldman Sachs has issued a business continuity statement that announces its preparedness through the triage of personnel; redundancy in servers and data infrastructures; dispersal across sites for archiving and service functions; and emergency electrical and building management systems, to name a few of their measures.[62]

The BCM practices of Goldman Sachs and other companies are marked by the absence of the state, and the geographic dispersal of these strategies across multiple national borders. Resilience continuity management demands the dispersal of activities into different zones of legal regulation, weather, energy, and geological stability to ensure the safety of its information networks and services. This is quite distinct from the spatial centralization of the Cold War bunker system. The continuity strategies of Goldman Sachs and other companies underscore a shift to the mode of vital systems security that has become dominant since the early 2000s and which is characterized by a disinterest in causal prediction in favor of all-purpose strategies of ensuring continuity—and, whenever possible, using disasters as occasions for further learning about how the corporate structure can better manage future disasters.[63] Resilience thus also participates in that zonal reformulation of territory in the name of experimentation that we described in chapter 2.

The recent concept of business *antifragility* developed by popular financial writer and sage of uncertainty Nassim Taleb is another example of adaptive management and resilience. While Taleb criticizes the specific way in which the term "resilience" has been employed by economists, his notion of antifragility is remarkably similar to the concept of ecological resilience developed by Holling. For Taleb, antifragility names the ability of organizations and organisms to gain strength from stress and shock. Taleb suggests that the awkwardness of his term "antifragile" underscores that there is no existing word for this concept in most languages. He claims, though, that this concept describes a dynamic that we can observe in natural evolutionary processes:

The most interesting aspect of evolution is that it only works because of its antifragility; it is in love with stressors, randomness, uncertainty, and disorder—while individual organisms are relatively fragile, the gene pool takes advantage of shocks to enhance its fitness. So from this we can see that there is a tension between nature and individual organisms.[64]

Evolution is radically uncertain. However, like ecologists and Goldman Sachs, adaptive managers can understand and produce new measures. As Taleb explains, antifragile managers—or, as we would put it, managers oriented toward resilience—constantly and actively measure error:

Fragility can be measured; risk is not measurable (outside of casinos or the minds of people who call themselves "risk experts"). This provides a solution to what I've called the Black Swan problem—the impossibility of calculating the risks of consequential rare events and predicting their occurrence. Sensitivity to harm from volatility is tractable, more so than forecasting the event that would cause the harm. So we propose to stand our current approaches to prediction, prognostication, and risk management on their heads. In every domain or area of application, we propose rules for moving from the fragile toward the antifragile, through reduction of fragility or harnessing antifragility. And we can almost always detect antifragility (and fragility) using a simple test of asymmetry: anything that has more upside than downside from random events (or certain shocks) is antifragile; the reverse is fragile.[65]

Taleb suggests that stress tests, failures, and simulations do not necessarily discover specific future risks but rather enable us to assess, and mitigate, systemic fragilities. As in the case of Holling's adaptive management and Goldman Sachs's business continuity plans, big data operate in Taleb's antifragility planning not in the service of imagining and instantiating specific futures but rather in consolidating and extending aspects of the present into the future.

Excursus 4.1
The East Kolkata Wetlands

The megacity of Kolkata, West Bengal, India, lies on the floodplains of the Hooghly River at sea level (figure 4.10). One of the largest and densest settlements on earth, the city was central to the development of capitalism and has long been at the heart of global trade and commerce. Kolkata was also a massive terraforming project for imperial and capitalist concerns. In search of wealth from saltpeter, opium, salt, silk, cotton, jute, tea, and rice, the British East India Company first transformed the silt and protective wetlands of the area into a logistical center through the creation of this riverine port in the late eighteenth century.[66]

This transformation of nature into real estate also inaugurated one of the world's great pandemics. Cholera bacteria lived untouched and untroubled inside small snails at the bottom of the river. When the East India Company's dredges uprooted these snails, and with them the bacteria they sheltered, they

4.10 Rajarhat, district 5, New Town, Kolkata, India. *Source*: Photo by Orit Halpern, March 13, 2016.

4.11 East Kolkata wetlands. (a) Fish farming. (b) Real estate speculation and high-end luxury development. *Source*: Photos by Sudipto Basu, October 4, 2021.

unleashed cholera into human populations. By the early nineteenth century, the first cholera outbreaks were reported in Europe. The pandemic continued for decades, becoming one of the most lethal pathogens afflicting human populations.

The wetlands of West Bengal have in this way long served as both sites of speculation and ecological catastrophe (figures 4.11a and 4.11b). Today, these silts and wetlands are rapidly being converted into shiny office towers in the search for smartness. The fancy peri-urban new and smart cities of the region, such as Rajarhat, Salt Lake City, and New Town, all tout slogans of a clean "atmosphere," of green construction, green corridors, and smart services and buildings. New Town, one of the shinier greenfield suburbs being constructed, has even recently been ordained by the Ministry of Housing and Urban Affairs as the "eighth most intelligent" of India's 100 smart cities. It has been certified and identified as "enabled" by an instrument labeled the *data maturity assessment framework 2.0* that measures how much data is put online.[67]

Bengali state planners hope for growth and innovation in Kolkata's technology sectors, which is lagging in comparison with other regions in India, by means of new developments on the former wetlands. While much of the new housing in Rajarhat, as well as in its nearby New City and Salt Lake developments, is underoccupied (having been bought for speculation by domestic and foreign investors), construction continues ahead at full speed on luxury condos and office parks. These spaces are highly leveraged and derived, conjoining joint imaginaries of smart and greenfield cities fostered by the Indian government and Bengali planners with derivative actions of global finance capital.[68] Whether these aspirations for smartness are realized or not, the developments, including a new Trump Tower, are commencing at a frantic pace. At the same time, some 30,000 people (as of 2016; the current total is surely much higher) have been displaced by the past decade of development.[69]

This search for real estate to build smart cities is, however, happening on top of the city's alternative infrastructures for resilience and life. Kolkata is perhaps the only city of its size on earth that has no wastewater treatment facilities. In the 1980s, sanitation engineer Dhrubajyoti Ghosh was asked by a finance minister to investigate the sewage treatment situation in Kolkata. He realized there was none. Instead, he discovered something long known to local fishermen and denizens (not to mention the millions eating the fish)—namely, that a system of aquatic agriculture dependent upon the local algal ecology was conducive to degrading wastes and could serve multiple functions. Seemingly miraculously, a city of 14.5 million inhabitants, 70 percent of whom live in poverty, has its water fairly effectively recycled and cleaned by a wastewater aquaculture area called the East Kolkata Wetlands (EKW).

The EKW help feed a city in which the majority of the population lives in poverty, provide cheap and energy-free sewage treatment, and—equally important—serve as one of the most important ecosystems and ecologies for flood defense in the face of climate change. The Rajarhat area in the wetlands also contains aquifers crucial for the water recharge of the entire delta and may maintain the hydrological equilibrium of large amounts of the subcontinental watershed. It is also a biodiversity hot zone.

The demise of the EKW to make way for smart buildings and infrastructure puts at risk the lives of the poorer denizens of Kolkata, as well as the existence of many nonhuman species. Environmentalists and planners have drawn on the concept of resilience to argue that destroying this area may be the death knell for a city that is one of the most vulnerable in the world to sea-level rise.[70]

Ghosh himself described forms of planning in which consideration of the poor is excluded from urban planning and engineering schemes as "cognitive apartheid."[71] In the past 30 years, many Asian cities turned from wastewater aquaculture to other forms of monoculture agriculture and abandoned wetland methods in the name of higher-technology sewage treatment plants. Kolkata had remained one of the exceptions to this rule. Though in 2002 the EKW was designated a Ramsar site and so is in theory protected under the purview of the United Nations Educational, Scientific, and Cultural Organization, ongoing dispossession—largely achieved through fraudulent means as poor fishing families are bought off—has transformed protected lands into speculative real estate. At the same time, the government touts sustainability and resilience primarily by understanding cities as "engines of growth for the economy . . . setting in motion a virtuous cycle of growth and development."[72] These contemporary practices extend earlier postcolonial policies encouraging urbanization and, supposedly, modernization, with smart cities as the next step in this process.

The irony, of course, is that just as the wetlands and their denizens are dispossessed and destroyed in places like Kolkata, wetlands have become central to urban planning in the rich Global North. In cities like Boston, New York, and Miami, the centrality of such designs is becoming increasingly understood as a key feature of smartness (figure 4.12). In these sites, ecosystems are understood to be acting as sensors, data collectors, filters, and nurseries for biodiversity as they protect the surrounding city.

Wetlands in this way highlight both the promise and the peril of smartness. Smartness produces resilience—but for whom or what is such resilience an issue? Smartness reimagines the divides between urban-rural and the hinterlands and the metropolis and creates new understandings of intelligence and knowledge. But it can also be recuperated to justify the destruction and sacrifice of certain lives at the cost of others. The Indian government, in its own search for smartness, has defined this concept broadly: since "there is no universally accepted definition of Smart City" then "the objective [of the smartness initiative] is to promote cities that provide core infrastructure and give a decent quality of life."[73] We agree, but note that in the absence of history and genealogy, the discourses of ecology and resilience are likely to be subsumed within neoliberal economy and environmental management.

4.12   Floating wetlands, Charles River, Cambridge, MA. *Source:* Photo by Orit Halpern, April 17, 2021.

RESILIENCE REPURPOSED

Each of our earlier examples—adaptive management, BCM practices, and antifragility strategies—instantiates in its own specific ways the more general logic by which smartness combines data gathering, modeling, and planning. Smartness predicates itself on a world assumed to be so complex that it can never be perfectly modeled, which in turn means both that catastrophes must be habitually expected and that one cannot plan for the next *specific* disaster. The logic of smartness thus endorses Paul Virilio's suggestion that we live in the era of the "generalized accident"—that is, an age characterized by catastrophes that are largely human induced yet nevertheless cannot be anticipated in their specifics.[74] The best one can do is to *manage* specific accidents as they arrive. Such management requires ever-increasing data collection within microenvironments, which is then networked in order to avoid the errors of centralized command and control. Within smart city logic, for example, the emphasis is on data-driven planning that can rapidly accommodate climatic, economic, or political change without being burdened by any specific political plans or dynamics.

At the same time, this effort to manage unthinkable futures encourages a constant search for "the long tail" (data that points to extreme and unusual events) since the very logic of planning is understood as inseparable from the temptation to take the norm—that is, what usually happens—for the entire space of possibilities. The assumption that catastrophe must be understood as habitual, so to speak, drives efforts to build technologies that can disperse risks in time and space, whether through the actual construction of diffuse and decentered physical networks or through forms of financial risk management, such as insurance or financial derivatives. All of these technologies require increased environmental calculation and computation, yet this increase of data gathering and technologies is not grounded in a discourse of certainty, causality, or positivism. Donald Rumsfeld, secretary of defense under President George Bush, perhaps put it best in his now infamous dictum that "there are known knowns; there are known unknowns; and there are unknown unknowns." Resilience planning is the means through which systems can prepare for, and (after the fact) learn from, those future unknown unknowns—all of which are, it goes without saying, assumed to be threatening and negative.

Returning to our opening example of COVID-19 curve flattening, these elements of smartness are at the heart of the strategies that governments have used to manage a pandemic that was unanticipated in its specifics but that was also, as we noted, anticipated by public health experts as a possible and even likely future scenario. A search online for "resilience" and "COVID-19" reveals a massive number of articles, websites, and consulting services dedicated to logistics, psychology, and community activism.[75] For managers of supply chains and corporations such as SAP and IBM, corporations must become resilient to ensure business continuity: "just-in-time" manufacturing has now become "just-in-case" manufacturing, and companies are urged to increase their options, to diversify supply chains geographically, to begin thinking about plasticity in manufacturing infrastructure (for example, being able to make alternative products), and to identify vital services and processes ahead of time. The COVID-19 pandemic has also clarified which populations national governments consider "expendable"—generally, the elderly, people with underlying health conditions, and people of color—as politicians seek to contain the spread of infection while at the same time ensuring continuity of the economy. As was especially evident in the response of the Trump administration in the US, resilience can be employed as a means of naturalizing violence by exploiting the uncertainties around a catastrophe—uncertainties, for example, about the precise vectors of transmission (surfaces and airborne respiratory droplets? Droplets only in closed rooms? and so on)—as a rationale for doing nothing (which in fact means allowing disproportionately high numbers of certain populations to die).[76]

Perhaps counterintuitively, though, resilience has also been invoked as a strategy, norm, and aspiration by some of those same groups positioned as expendable within government and corporate strategies of dealing with COVID-19. On the Black Lives Matter website, for example, resilience is imagined as an alternative to the triage logic of the status quo: "We affirm our humanity, our contributions to this society, and our resilience in the face of deadly oppression." Critical race theorist Kara Keeling has drawn on Taleb's related concept of antifragility as a figure of thought for Black liberation and for the possibility of becoming stronger through exposure to ongoing shock. Keeling stresses that Taleb's book itself is a neoliberal treatise and is interested only in the implications of antifragility for commercial entrepreneurialism. Yet she notes that precisely because Taleb's concept of antifragility is a *critique* of the efforts of economists to predict and control future risks through computational and calculative techniques of derivation and commensurability, it can be part of a liberatory strategy. More specifically, Keeling suggests that

the concept of "antifragility" offers the following to the present project [of her book *Queer Times, Black Futures*] and its investments in freedom dreams: (1) a critique of finance capital's construction of "futures" . . . and (2) another way of thinking about the queerness in time as an ally in building the antifragility of freedom dreams, the obsessive love that sustains them, and those who advance such dreams within, without, and through love.[77]

The concept of antifragility is useful to Keeling—and the related concept of resilience is presumably useful to the authors of the Black Lives Matter website—because these concepts focus attention on the connection

between thriving and shock. Keeling suggests that "something is antifragile when it thrives rather than breaks in conditions of disorder and
randomness," and hence "Taleb's work enables the insight that Black
cultures are antifragile. They build accidents and surprises into the mod-
ulations that enable them to endure."[78] For Keeling, the concept of antifragility calls for us to understand the future of our present as not yet
decided and so as potentially radically different from the present. Keeling's concept of shock thus does not legitimate the sacrifice of lives but
rather recognizes that trauma has been ongoing and continuous for Black
people and many others. Such trauma, however, not only can be survived
but can become a source for creativity and transformation.

Excursus 4.2
Possible Futures of the Smart Forest

In ecology, too, resilience is now a contested concept and tool for reimagining
more diverse futures and forms of life. Suzanne Simard, a forestry professor at the
University of British Columbia, is a leading voice in contemporary forest management and is best known for her work on forests as communication systems
that have "social lives."[79] In the late 1980s and 1990s, Simard (like Holling a
few decades earlier) worked for the Canadian forestry industry. Her studies demonstrated, again, that when loggers replaced diverse forests with homogeneous
plantations the new trees failed to thrive. However, Simard, perhaps unintui-
tively, noted that part of this plantation system required the clearing of underbrush and the movement of soil, so she began to examine the soil. Her article
and thesis in the mid-1990s on the subject changed the field significantly.[80]

Simard discovered that trees communicate through networks of mycorrhizas
that exchange carbon, phosphates, and hormones. She subsequently employed
genomics, radiation sensing, and a combination of big data from satellite, forest sensing, and genomic testing to illuminate the incredible complexity of
biodiversity and its importance for conservation. A central component of her
approach was the idea of a "mother tree": a living archive of both knowledge
and energy for the forest. These trees foster seedlings and are critical nodes that
permit the forest to thrive. Advanced mapping of this network—which was, in
essence, a smart or big-data project, though for unusual ends—facilitated an
understanding of the key infrastructures that, if destroyed, lead rapidly to the
demise of the entire ecosystem.[81]

The work of Simard and fellow ecologists creates a different visualization
of the forest (see figures 4.13 and 4.14). Major news outlets translated this

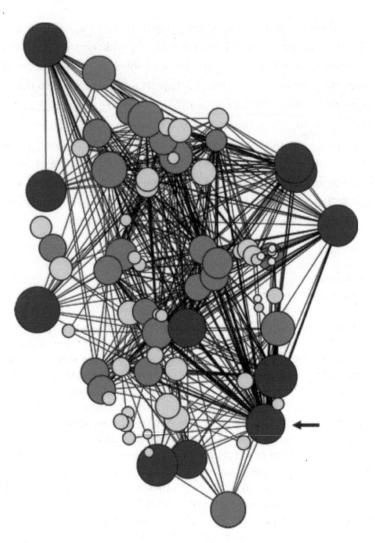

4.13   Network image showing the linkages between Douglas fir trees through the mycorrhizal network. The arrow points to the most highly connected tree. *Source*: Beiler et al. 2010, displayed online in Suzanne W. Simard, "Nature's Internet: How Trees Talk to Each Other in a Healthy Forest," TEDxSeattle, February 2, 2017, https://www.youtube.com/watch?v=breDQqrkikM.

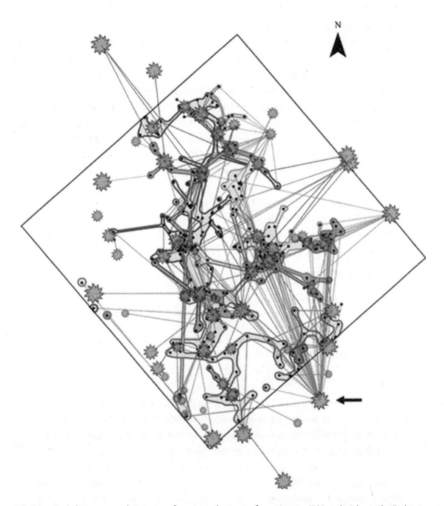

4.14 Another network image of tree exchange of nutrients. "Woodwide webs" showing links between older "mother" trees, saplings, and other species of trees. The lines illustrate the flow of chemicals and energy between the trees. The studies even demonstrated that trees can signal each other to begin preparing against predators, such as budworms. This incites the surrounding trees to begin preparing defenses (sap) against possible insect invasion. *Source*: Suzanne W. Simard, "Nature's Internet: How Trees Talk to Each Other in a Healthy Forest," TEDxSeattle, February 2, 2017, https://www.youtube .com/watch?v=breDQqrkikM.

data and approach into the claim that forests have a "social life," that trees "communicate" with each other, and that we should never "underestimate the intelligence of trees."[82] Simard herself, however, discusses the forest in terms of smartness and calls a forest "an internet of trees"; in this way, she clearly asks her audiences to understand a forest in terms of the communication networks usually assigned to digital media. And for Simard this is not simply a metaphor, for her modes of visualization—from Geiger counters and tracing the spread of radiation among mycorrhizal networks and through root systems and soil to genomic analysis—are all forms of digital visualization that became possible in the past recent decades and facilitated the production of models that could not be seen earlier (and hence could not be imagined).

Simard cites Gunderson and Holling as her predecessors and extends their understanding of biodiversity into adaptation. As she and her colleagues write in a recent summary article, they understand adaptation to be the next step in understanding ecosystems:

The adaptive capacity of many complex systems is related to the concept of ecosystem resilience [developed by Holling and Gunderson] but with an important difference. Ecological resilience can be characterized by the amount of change that an ecosystem can absorb before it loses its ability to maintain its original function and structure, i.e., its identity. Following a disturbance, a resilient system has the potential to recover its original structure, functions and feedbacks. In contrast, adaptation enables an ecosystem to modify its structure and composition so it can sustain major functions or develop new ones. It enables the ecosystem to reorganize in a manner that avoids maladaptation to the new environmental conditions.[83]

Within the new matrix of ecology governed by means of complexity theory, true resilience is no longer simply sustainability but adaptability. This is a form of adaptability, moreover, that only comes to be through networks of smartness—that is, codependencies and relations between many parts of the forest and diversity in the underbrush, in the trees, and in the soil.

Mycorrhizal networks are essential for this approach, for they share chemical signals, nutrients, and carbon among the trees. These forms of communication, sharing, and signaling play critical roles in adaptation. These networks, ecologists argue, are not only information systems like markets, signaling different life forms to adapt. The networked forest, ecologists argue, also has a *memory*:

Complex systems accumulate information from the past that influences future trajectories through persistent change in the system's structure and composition. . . . This *memory* may derive from past events, some minor or random, that are reinforced through feedbacks in the system and constrain its future trajectory. In forests, early recruitment of pioneer species following a disturbance modifies the habitat and influences prospective colonists. . . . Forest management practices may also create path dependency, as for example in Canada, where an emphasis on stand-level silvicultural planning still drives current management practices. This emphasis persists despite new technologies (such as

GIS and GPS) that readily permit management at larger scales that may be more ecologically and economically relevant.[84]

Forests have memories, and these memories shape the future of the ecosystem. These memories are held within the material bodies of the plants and related species, as well as in the choices made by humans and by the vagaries of which trees are grown and where. For Simard, the forest's ability to communicate, the strength and vitality of the nodes of mother trees, and the trees' connection to other species and their saplings are all questions of history. A system that has been devastated and monocropped will lack the material memory to adapt to future changes.

This memory augments the machine systems that both make this forest visible and enable it to be destroyed. The proto–geographic information and proto–global positioning systems that Holling first used in Canada allow for ongoing understanding of the changing densities, species, and temperatures of the forest. However, this knowledge must be used within the context of understanding the system as changing and always evolving, rather than as static.

This understanding of the forest and how one can maintain its "services" has been critical to changing the practices of forest management. This underscores the more general point that specific ways of visualizing life have material impacts on that life itself. Simard and her colleagues argue that good policies encourage forest harvesting that respects the network and avoids damaging critical nodes. They also argue that wood *can* be harvested, but only if care is taken not to clear cut and attention is paid to regrowing the forest undergrowth, maintaining the soil's diverse life forms, and reintroducing multiple tree species on the basis of an understanding of their interrelationships.

We have highlighted this new understanding of adaptability because it also highlights the positive potential of the concept of resilience. In this case, new forms of digital data and imaging, combined with new models of networks, learning, adaptation, and communication, have led to forms of managing forest futures in more diverse ways. Complexity systems thinking in forestry understands the need for change and understands that forests are always in transition while also emphasizing "holistic" management approaches, which (to paraphrase Holling) enhance adaptability and resilience for futures that cannot be fully predicted. Diversity is a value in part because of uncertainty.

CONCLUSION

At stake in the competing understandings of resilience and antifragility that we have noted above is the common question of what, precisely, "learning with continuity," to draw on Holling's powerful phrase, can and should mean and how computation fits into that aspiration. We can parse this into four questions: What is *learning*? What must *remain continuous* for learning to happen? What is the role of *computation* in learning with continuity? And on what *image of biological evolution* does such learning model itself?

We take up the second question first, for the question of what must *remain continuous* for learning to happen provides the clearest axis of distinction among the different images of resilience and antifragility that we have discussed in this chapter. BCM stands at one extreme. This practice focuses solely on the individual corporation and seeks to ensure that the core business—and ultimately the elements of the core business that affect the corporation's share price or profit margin—remains continuous. The environment of the business, whether understood as the natural, political, or social environment, is relevant only insofar as it bears upon the ability of the legal entity of the corporation to persist through time. In practice this means that the resilience at which BCM aims requires the persistence of the legal, political, and social status quo. The problem with that approach to resilience, Keeling suggests, is that it forecloses on a future not predicated upon racial capitalism. Drawing on the example of Royal Dutch Shell—the company that, as we noted above, was central to the development of scenario planning as a corporate practice—Keeling contends:

Royal Dutch Shell's existence is predicated on a system of racial capitalism that thrives on the dispossession and exploitation of Black people, Indigenous peoples (some of whom describe themselves as "Black"), and people of color. A future in which Royal Dutch Shell would continue to exist as such forecloses upon a future in which those groups of living beings we currently can identify as "Black people" and/or Indigenous peoples, have the resources to enjoy a sustainable and joyful existence on this planet.[85]

For Keeling, the learning that is necessary in order to create a future that is *not* simply an extension of the inequities of the present can be found in the continuous and resilient "freedom dreams" of what Cedric J. Robinson first called the Black radical tradition, which has thrived not simply in the face of, but because of, the shocks and trauma that partly engendered it.[86] Where corporate, financial, and logistical approaches to resilience assume a world of scenario plans and unanticipatable futures divorced from historical legacy or context—and where focusing on services eliminates any need to consider the specifics of environments or milieus—Keeling understands resilience as inextricably linked to an extended historical consciousness and to the conscious and planned redesigning of institutions and environments.

Keeling's stress on the link between resilience and an extended historical consciousness seems to us well worth endorsing. This link underscores the extent to which Hayek's image of the market as information processor—and, to a lesser extent, Holling's image of ecosystem services that persist even as individual species come and go—has no interest in history: so long as the market (Hayek) or ecosystem services (Holling) persist, the species of the past are of no interest or relevance. Hayek's and (sometimes) Holling's disinterest in history is drawn from their understanding of evolutionary biology, for whatever evolutionary links there might be between a species in the present and the ancient species to which it is related is of no interest or help to the present species in its struggle for survival.

This in turn highlights the difficulties of drawing on biological evolution as a model of *learning*. While biological speciation certainly involves *change*, it is hard to see what, precisely, *learns* as a consequence of that change (if only because no species gets a chance to learn from its mistakes after it has gone extinct). Keeling's approach thus suggests that understanding resilience as a call for multiplicity, and for futures not yet known, will require a mode of ecological thinking that runs counter to the optimizing demands of capital and that can offer the possibility not of a new normal but a new nature.

At the same time, we also need to rethink the role of computation and sensing (and the infrastructures and institutions that support these functions) in enabling this new ecology of learning. Computation, in Keeling's account, seems to be primarily a support of racial capitalism, and it is possible to imagine, following Fred Moten, that in the face of the smartness mandate or any other aspect of the normative present "the only thing we can do is tear this shit down completely and build something new."[87] If we return to our example of the COVID-19 pandemic, though, it is not entirely clear what this means or how it would differ from the current strategy of the political Right. Presumably, it *is* a good thing to track and seek to control COVID-19 curve rates via smart technologies, to analyze this data carefully to determine which communities are suffering disproportionately and to seek to ameliorate those differences, and to provide intensive care to anyone who needs it. The alternative strategy favored by the Right—namely, actively dismantling the systems that would allow for this kind of tracking and care—in the end amounts to a sort of economically oriented triage logic, according to which those who have the economic means ensure care for themselves, and everyone else is forced to hope for the best.

The strategy of tearing everything down also arguably overestimates the coherence of "the system" and so misses chances for both learning and continuity. The genealogical approach that we take in this book is intended to underscore the (generally unintentional) bricolage that has brought together the premises, technologies, and *dispositifs* of the smartness mandate. This bricolage in turn means that these networked computational technologies can be used to other ends. One starting point, for example, would be to identify those services—for example, equitable health care or universal food and emotional support during a global pandemic—that we wish to remain continuous and to figure out concretely the ways in which the tools of smartness can be reoriented toward those goals. In Simard's new ecology of the smart forest, we can perhaps see at least an inkling of what such a rethinking might mean. And in our coda, we will turn to what it might mean to explore the layered temporalities and bricolages of smartness.

CODA: FROM THE SMARTNESS MANDATE TO THE BIOPOLITICAL LEARNING CONSENSUS

The smartness mandate is the demand, cast by its advocates as having the force and irresistibility of a law of nature, that all social processes become smart. A social process becomes smart when the populations within which that process occurs are redesigned as experimental zones, so that widely distributed forms of electronic sensing produce data that can be processed algorithmically, and in this way enable constant and quasi-automatic learning about, and adaptation to, an ever-changing environment (i.e., resilience). Because smartness is presented as a—or more often, the only— means by which humans can successfully adapt their current highly technical collective existence to those threatening changes in global ecology and geology that humans themselves have engendered, the smartness mandate seems to emanate less from sites and structures of human government than from nature itself and its evolutionary processes.

As we have documented in this book, many of the guiding premises and models for the smartness mandate first emerged in what we might broadly call the cybernetic sciences of the 1940s and 1950s. These elements were linked to one another in new ways in the 1970s and 1980s, often as responses to new forms of global relations, including postcolonial movements (which called into question the post–World War II geopolitical order), changes in the structure of global monetary flows and finance (the collapse of the Bretton Woods international currency system enabled, for example, a new system of finance and derivatives), and ecological concerns (which stressed for many the inability of the traditional nation-state to address problems that extended beyond national territories). These new linkages solidified into the smartness mandate in the early 2000s, often guided by the convergence of neoliberal policies and the new capacities of computing and sensing technologies.

Yet we have also stressed in each chapter both the contingent nature of these connections and the alternative possibilities enabled by the premises, models, and techniques that make up the smartness mandate. In our coda we expand on these alternative possibilities, and we do so in the name of what we will call the *biopolitical learning consensus*. The smartness mandate seems to be a mandate in part because of the high stakes involved: for its advocates, we *must* become smart or else go extinct as a species. And the smartness mandate enjoins us to smartness—rather than, for example, rationality—in order to underscore the inability of unassisted human reason to understand and cope with the modern challenges that humans face; as a consequence of this incapacity, humans need learning processes that take place largely within computer systems and that have no telos other than perpetual resilience.

What we call the biopolitical learning consensus agrees that unassisted human reason cannot fully understand and cope with the modern challenges that humans face and that, as a consequence, humans need learning processes that take place at least partly within computer systems. Yet for the biopolitical learning consensus, the limits on unassisted human reason stem less from intrinsic limits of rationality than from the fact that there is no one group of humans that would ever be able to define the nature and contents of rationality or human reason. Rationality is *not* something that can be defined and axiomatized by, for example, the game theorists of the 1960s; rather, it is a capacity that is always at work and that works differently in every human collective. Learning processes that take place in part within computer systems thus need to remain open to different perspectives on both what constitutes rationality and what counts as learning.

In place of a *mandate* to be smart, the biopolitical learning *consensus* seeks to bring these different perspectives together—that is, to feel and think together with them. And unlike past examples of political "consensus" that in fact represented agreement among a very small number of people—for example, the neoliberal "Washington Consensus" that emerged in the late 1980s—this consensus has no geographic location but rather comes together through the distributed efforts of those interested in learning in all its forms.

The linkage we make between learning and plurality seeks, in one sense, to lead the thought of populations away from their conceptual and literal capture by markets. In chapter 1 we presented biologist Ernst Mayr and economist Friedrich Hayek as uncanny doubles of one another. Neither was interested in equilibrium states but rather in what happened when conditions *changed*. Both felt that differences between each of the members of large collectives—biological populations, for Mayr; market participants, for Hayek—were key to understanding both the capacity for, and the actual process of, adaptation to a changing environment. For both Mayr and Hayek, this process of adaptation was invisible to the participants within this process: a subspecies adapts to its changing environment by transforming into a new species, but no member of the species can be said to learn in this process. In similar fashion, the market "processes" the local information possessed by each market participant by means of the prices of goods, but neither the market as a whole nor any market participant learns, precisely. Yet, perhaps paradoxically, the points of conceptual resonance between Mayr's and Hayek's accounts enabled concepts from each of their respective disciplines to serve, beginning in the 1950s, as points of orientation for computer models of *human* learning and, subsequently, for models of *computer* learning.

For the biopolitical learning consensus, the concept of population is much more helpful than the concept of the market for reappropriating concepts of learning. This is in part because, even in Mayr's biological formulation, population has a much more capacious sense of difference than does Hayek's market. While Hayek's market participant is, like Mayr's biological individual, a site of difference, Hayek's market eliminates any difference that cannot be related to the market metric of price. While fitness might seem to play the same role in Mayr's account, in fact the persisting ambiguities of what, exactly, constitutes biological fitness necessarily make this a more capacious concept. However, beyond this narrow biological point, population is also a term employed within numerous discourses, including demography, public health, and public policy, which opens up this concept toward what Roberto Esposito and others have described as an *affirmative* biopolitics.

In each chapter we have sought to illuminate contemporary experiments that direct collective intelligence, sensing, computation, ecology, and economy toward more equitable shores—that is, toward what we are calling the biopolitical learning consensus. As a way of illuminating the latter even further, we close with some final reflections under two headings: the relationship of learning to democracy and the relationship of learning to history and memory.

LEARNING AND DEMOCRACY

Smartness is understood by its advocates as a method of perpetual learning. For this reason it is fundamentally distinguished from rationality, which in all of its modalities (ancient, classical, Cold War, and so on) focuses on stable criteria for judgments. Smart technologies are perpetually involved, of course, in making judgments. However, these judgments are understood as always provisional and error prone and hence primarily occasions for further learning. Moreover, both these provisional judgments and the open-ended process of learning that they enable cannot be restricted to the conscious decisions of humans because they also involve *automated* judgments based on large data sets and environmental systems of sensing. And because smartness presumes an environment that can never be fully known or mapped, learning has no endpoint: that is, there can be no state of knowledge or decision-making beyond smartness.

This approach to judgment and learning sets up a potential conflict with the values and procedures of democracy, though this differs in important ways from the tension between democracy and neoliberalism. As historians of economics such as Philip Mirowski and Edward Nik-Khah have documented, neoliberal economists—who often also served as shock doctrine consultants—made no secret of their disdain for democracy whenever the latter seemed to question the purported wisdom enabled by markets.[1] For neoliberal economists such as Friedrich Hayek, George Stigler, and Milton Friedman, the process of voting that occurs within a representative democracy is a coarse and problematic approximation to the distributed decision-making that occurs through purchasing decisions; ideally, then, the decisions currently made by means of democratic political processes would be shifted into purchasing decisions.[2]

On the one hand, smartness is also characterized by an elective affinity to markets, since the latter employ a quantitative measure (price) that can easily be integrated into smart learning algorithms for the purpose of "weighting" elements of the latter. On the other hand, smartness is *not* committed to price as the measure of all things, for the "wisdom" of its distributed populations can also employ nonprice metrics (e.g., page-ranking methods, Wikipedia, etc.). Yet because its distributed knowledge can *only* be harnessed by means of environmental sensing and algorithmic computation, smartness still bears a fundamentally ambivalent relationship to democracy, since it is not clear what role machines and automated modes of decision-making ought to play in democratic processes.

We wonder whether democracy and smartness might be brought closer together by thinking about the latter through the example of one of the supposed detriments of liberal democracy—namely, its tendency toward perpetual conversation. Twentieth-century critics of democracy such as Carl Schmitt and Leo Strauss pointed to what they saw as a contradiction between the democratic assumption that the diverse perspectives of the populace must be respected in any concrete decision and the fact that every concrete decision necessarily negated the validity of the perspectives of those who voted against the decision.[3] For Schmitt and Strauss, this meant that democracy was either a perpetual conversation that avoided ever arriving at concrete decisions or that it must make recourse to extrademocratic states of exception in order to make concrete decisions. Yet the temporal orientation of smart learning—its open-ended commitment to revision and change—might itself serve as a figure for the open-endedness of democracy, in the sense that smartness necessarily involves decisions but does not necessarily consider any of those final. From this perspective, the fact that the quantitative dimension of smartness is not bound to the market (even if it often finds itself drawn into the gravity of the latter) certainly makes it a better bedfellow for democracy than neoliberalism proper, for the latter precludes any court of appeal beyond the market's red tooth and claws. (Nor, we will add, are the aspirations of contemporary democracy necessarily alien to big data and algorithms, for it was precisely the latter that enabled critics of gerrymandering to prove to the US Supreme Court that such had in fact taken place in North Carolina in the United States.[4])

The question of what learning means is clearly key to any effort to bring smartness and democracy closer to one another. If smart learning is understood through a Darwinian image of winners ruthlessly divided from losers in a process of evolution, smartness will likely always find itself unable to escape the event horizon of the market. We have pointed, however, to other images of natural evolution—for example, the smart forest—that can move us toward other images of learning. Nor does learning itself need to be bound to a biological image. To return to our opening example of the Event Horizon Telescope, we find hope in the signals of a black hole, which were sent eons ago from a time beyond human—even Terran—time. This is a reminder that some experiences can only emerge through the global networks of sensory and measuring instrumentation. There are in this sense radical possibilities in realizing that learning and experience might be not only internal to subjects but also shared. Perhaps these are just realizations of what we have known all along: that our worlds are composed of relationships to Others.

HISTORY AND MEMORY

As we have noted at multiple points in this book, the "success" of smartness is in large part a function of its agnostic relationship to crisis and catastrophes. Rather than seeking to specify in advance what might constitute a crisis—a specification that would itself require explicit value judgments—smartness instead presents itself as an all-purpose method for responding to any and all crises, whether these seem primarily political (e.g., terrorism), economic (e.g., collapse of futures markets), medical (e.g., pandemics), or ecological (e.g., rising sea levels). Smartness is both an epistemology and a set of technologies for ensuring consistency throughout multiple crises, or what Holling called "learning with continuity." Though smartness has a memory of sorts, in the sense that each crisis should be an occasion to learn even more about the means for ensuring consistency in the next crisis, there is no space in the epistemology of smartness for history, memory, and historical trauma in the sense in which we usually understand those terms.

We have thus sought, at various points, to open up smartness to traditional understandings of history, memory, and historical trauma. In some cases this was via an appeal to alternate understandings of natural evolution: for example, Simard's understanding of the smartness of forests as in part a function of their capacity for extended memory. In other cases we emphasized the long lineages of trauma of those deemed "losers" by a market-oriented vision of smartness (for example, those who lost houses in the global market crash of 2008, who were overwhelmingly also those who would earlier have been denied housing loans due to racial redlining policies). In addition, we excavated alternative visions of computation and living, such as the Japanese architect Arata Isozaki's City in the Air project, which was like, yet differed significantly from, Negroponte's vision of soft architecture. Had these past experiments been embraced, they would have led smartness in different directions. They are in this sense especially important to our project, for they underscore that the smartness mandate was not inevitable but instead resulted from contingent connections and alliances.

To stress even further this fact of contingency, we wish to recall that although machine learning since the 1950s has become a project of revising cognition, it was originally an attempt to understand better what we might call the historical dimension of human perception. The first model of a neural network, for example—Frank Rosenblatt's perceptron—was not introduced as a model of artificial intelligence but was intended to teach us about natural intelligence. The perceptron was not a technical solution to a problem (how to automate pattern recognition) but rather a way to pose questions and learn about an unknown entity (natural intelligence). Or, more accurately, it was an experimental construct for producing models, which asked: How might such a model change how we understand what a machine or a mind *might* be?[5]

One answer to that question emerged from the genealogical relationships among Rosenblatt, Hayek, and Hebbs (as well as Donald Ewen Cameron, Milton Friedman, Augusto Pinochet, and a host of others) and eventually gave rise to what Naomi Klein called the shock doctrine. Yet the movement from the shock doctrine to the smartness mandate reveals yet a different answer, or set of answers, to the question of what machines and minds—and their interrelationships—might be. Our hope, of course, is that this book can, by returning us to this contingent genealogy, help us to formulate yet other answers.

We conclude by returning to the beginning. The Atacama Desert, where we began our prologue, is, on the one hand, the site of new capacities: a site, for example, in which astrobiologists can locate new forms and modes of life; a site for which computer scientists can develop new mathematics of real-time monitoring and modeling of massive copper and gold mines; and a site by means of which astronomers can produce new images of aspects of the universe, such as the event horizon of a black hole. On the other hand, the production of these new capacities in the Atacama also seems to bring its own forms of death by contributing, for example, to the disappearance of native flora and fauna, including Indigenous human groups. But even in the face of this apparent loss and death, smartness is positioned by its advocates as a potential savior. Sociedad Química y Minera engineers, for example, assert that new technologies will allow water managers to optimize water usage by recycling and collecting the water that evaporates. Smartness purportedly enables, in other words, the *most* finite resource in the desert (water) to become elastic and optimizable and allows the environment more generally to become fortified and made resilient. From this perspective, resource limitations and catastrophic environmental events no longer emerge as "crises" that necessitate responses by experts (even those economists who are experts in "austerity" fiscal policies). They can instead be addressed through incremental experiments, which enable would-be crises to be ameliorated through endless adjustments and manipulations of time and data collection.

Yet time and data can be manipulated in many ways, and one specific manipulation of this same desert—Patricio Guzman's film *Waiting for the Light* (2010)—underscores some of what is lost in these dreams of smartness. In the immediate aftermath of the September 11, 1973, coup in Chile, nearly 10 percent of the national population was tortured, "disappeared," or exiled. Some of those who were disappeared ended up in the desert, first taken by Puma helicopter from detention sites and then either killed and buried in unmarked graves or thrown from the helicopter. These disappearances were only a small part of a larger program of torture and murder. Over 2,000 Chileans were murdered, and tens of thousands tortured. Thousands also fled the country (see figure C.1).[6]

C.1 Calama Memorial for the Pinochet victims. *Source*: Wikimedia Commons, accessed August 6, 2019, https://commons.wikimedia.org/wiki/File:Memorial_DDHH_Chile_06 _Memorial_en_Calama.jpg.

Guzman's film creates visual parallels between, on the one hand, the search by mothers for the bodies of their children killed by Pinochet's supporters and, on the other, astronomers watching and recording the stars in the Atacama's high-altitude observatories. (The wave millimeter arrays were not yet functioning when Guzman created his film). Guzman treats the desert landscape as a recording machine for both human and inhuman memories: the search for murdered loved ones and the trace of stars 50 million light-years away. The film thus suggests that the desert provides some kind of intelligence and memory that is partially accessible, but not restricted, to humans. Guzman offers us a dangerous romance with the possibilities afforded by our nonhuman intelligences that augment or supplant our human memories (see figures C.2 and C.3).

C.2   Patricio Guzman, *Waiting for the Light*, 2010.

C.3   Patricio Guzman, *Waiting for the Light*, 2010.

His story ties together tales of cybernetics, astronomy, and economy in a way that confronts both the horror of and our capacity to imagine encounters with radical forms of difference. In many ways it refracts the event horizon as an image. It is the image of what we cannot see. It is proof of the limits of scientific representation. Our hope is that by reflexively encountering through our machines the very limits to human knowledge and control, we might envision another path for smartness—a path that, by recognizing the limits to computation, realizes new possible relations of care and participation to the world.

NOTES

PROLOGUE

1. Wikipedia, s.v. "event horizon," last modified January 29, 2022, https://en .wikipedia.org/wiki/Event_horizon.

2. Dennis Overbye, "Darkness Visible, Finally: Astronomers Capture First Ever Image of a Black Hole," *New York Times*, April 10, 2019, https://www.nytimes.com /2019/04/10/science/black-hole-picture.html.

3. Another technique employed here is to cool units at the base of the telescopes to the temperature of deep space in an effort to isolate and process signals from space and separate them from "noise" from the earthly atmosphere. By returning the signal to its original temperature, the appropriate wavelengths of the signals can be isolated. In this installation, then, data is literally being contextualized in an environment that is built within the experimental setup, in the sense that the stellar "outside" of the earth is recreated within the machines. This was explained to Orit Halpern by an on-site technician on March 13, 2021.

4. Martîn Arboleda, *Planetary Mine: Territories of Extraction under Late Capitalism* (New York: Verso, 2020), 66.

5. Staff reporter, "Codelco to Deploy AI Solution," *Mining Journal*, March 26, 2019, https://www.mining-journal.com/innovation/news/1359598/codelco-to-deploy-ai -solution.

6. Lithium was first discovered in 1817 by Swedish chemist Johan August Arfwedson. Arfwedson, however, was not able to isolate the metal when he realized that petalite contained an unknown element. In 1855, British chemist Augustus Matthiessen and German chemist Robert Bunsen were successful in separating it. It is one of the lightest and softest metals. In fact, it can be cut with a knife. And because of its low

density, lithium can even float in water. Terence Bell, "An Overview of Commercial Lithium Production," Thought Co., last modified August 21, 2020, https://www .thoughtco.com/lithium-production-2340123; "Lithium," Royal Society of Chemistry Periodic Table, accessed March 15, 2021, https://www.rsc.org/periodic-table/element /3/lithium.

7. Alejandro Bucher (technical manager, SQM), interview by Orit Halpern, March 23, 2017. See also Paul Harris, "Chile Seawater Desalination to Grow 156%," *Mining Journal*, January 27, 2020, https://www.mining-journal.com/copper-news/news/1379729 /chile-seawater-desalination-to-grow-156; Rebecca Boyle, "The Search for Alien Life Begins in Earth's Oldest Desert," *Atlantic*, November 28, 2018, https://www.theatlantic .com/science/archive/2018/11/searching-life-martian-landscape/576628/.

8. Michelle Carrere, "Chile Renews Contract with Lithium Company Criticized for Damaging Wetland," trans. Sydney Sims, *Mongabay*, December 16, 2018, https:// news.mongabay.com/2018/12/chile-renews-contract-with-lithium-company -criticized-for-damaging-wetland/.

9. Lorena Gúzman, "The Fight for the Control of Chile's Lithium Business," *Diálogo Chino*, December 7, 2018, https://dialogochino.net/15614-the-fight-for-control-of-chiles -lithium-business/.

10. Carrere, "Chile Renews Contract."

11. Frank Tavares, ed., "Cooking Up the World's Driest Desert—Atacama Rover Astrobiology Drilling Studies," NASA, June 20, 2018, https://www.nasa.gov/image -feature/ames/cooking-up-the-world-s-driest-desert-atacama-rover-astrobiology -drilling-studies.

INTRODUCTION

1. Sam Palmisano, "A Smarter Planet: The Next Leadership Agenda," Council on Foreign Relations, November 6, 2008, https://www.youtube.com/watch?v=i_j4-Fm_Svs.

2. Palmisano, "A Smarter Planet."

3. Tom Warren, "Microsoft Teams Jumps 70 Percent to 75 Million Daily Active Users," *The Verge*, April 29, 2020, https://www.theverge.com/2020/4/29/21241972/micro soft-teams-75-million-daily-active-users-stats.

4. Paul Erickson, Judy L. Klein, Lorraine Daston, Rebecca M. Lemov, Thomas Sturm, and Michael D. Gordi, *How Reason Almost Lost Its Mind: The Strange Career of Cold War Rationality* (Chicago: University of Chicago Press, 2015), 2. Erickson and his coauthors stress that for Cold War authors and policy-makers, the possibility of nuclear war made it imperative that people—or at least military commanders and policy-makers—act "rationally," in the sense that tendencies to innovate or depart from programmable rules be prevented; the consequence was that "mechanical rule following . . . become the core of rationality" (31).

5. Though the image of Cold War rationality developed by Erikson et al. is especially useful for our purposes here, we also want to acknowledge alternative histories of temporality and control, many emerging from cybernetics, within the history of

Cold War computing. See, e.g., Orit Halpern, "Cybernetic Rationality," *Distinktion: Scandinavian Journal of Social Theory* 15, no. 2 (2014): 223–238.

6. Naomi Klein, *The Shock Doctrine: The Rise of Disaster Capitalism* (New York: Metropolitan Books/Henry Holt, 2007). Klein's book is part of an extensive bibliography of recent critical work on neoliberalism that also includes David Harvey, *A Brief History of Neoliberalism* (New York: Oxford University Press, 2005); Philip Mirowski and Dieter Plehwe, eds., *The Road from Mont Pèlerin: The Making of the Neoliberal Thought Collective* (Cambridge, MA: Harvard University Press, 2009); Jamie Peck, *Constructions of Neoliberal Reason* (New York: Oxford University Press, 2010); and Philip Mirowski, *Never Let a Serious Crisis Go to Waste: How Neoliberalism Survived the Financial Meltdown* (New York: Verso, 2014).

7. See especially Michel Foucault, *The History of Sexuality*, vol. 1, *An Introduction*, trans. Robert Hurley (New York: Pantheon Books, 1978); Foucault, *Society Must Be Defended: Lectures at the Collège de France, 1975–76*, ed. Mauro Bertani and Alessandro Fontana, trans. D. Macey (New York: Picador, 2003); Foucault, *Security, Territory, Population: Lectures at the Collège de France, 1977–78*, ed. M. Senellart, trans. G. Burchell (New York: Palgrave Macmillan, 2007); and Foucault, *The Birth of Biopolitics: Lectures at the Collège de France, 1978–79*, ed. M. Senellart, trans. G. Burchell (New York: Palgrave Macmillan, 2008); Gilles Deleuze, "Postscript on the Societies of Control," *October* 59 (1992): 3–7; and reflections on immaterial labor in Maurizio Lazzarato, "Immaterial Labor," in *Radical Thought in Italy*, ed. Paolo Virno and Michael Hardt (Minneapolis: University of Minnesota Press, 1996), 132–146; and Michael Hardt and Antonio Negri, *Empire* (Cambridge, MA: Harvard University Press, 2000), 290–294.

8. For a key early reflection on biological "population thinking," see Ernst Mayr, "Darwin and the Evolutionary Theory in Biology," in *Evolution and Anthropology: A Centennial Appraisal*, ed. B. J. Meggers (Washington, DC: Anthropological Society of Washington, 1959), 1–10. For a helpful reflection on key aspects of biological concepts of population, see Peter Godfrey-Smith, *Darwinian Populations and Natural Selection* (Oxford: Oxford University Press, 2009).

9. On complicated and shifting relationships between natural and social scientific approaches to population in the twentieth century, see Edmund Ramsden, "Eugenics from the New Deal to the Great Society: Genetics, Demography and Population Quality," *Studies in History and Philosophy of Biological and Biomedical Sciences* 9 (2008): 391–406.

10. See "Recommending for the World," *Netflix Technology Blog*, accessed October 27, 2016, https://netflixtechblog.com/recommending-for-the-world-8da8cbcf051b.

11. As these examples suggest, we see the concept of population as more useful for an analysis and critique of smartness than contemporary alternatives such as crowds, swarms, and collectives. While each of these terms admittedly stresses different aspects—population emphasizes long-term biological adaptability and persistence, crowds and swarms emphasize speed of change and decentralized control, and collective is clearly a more political term—the concept of population underscores the evolutionary logic of smartness, as well as the underlining meanings of optimization

and resilience central to its operation. The concept of "multitude" employed (in different ways) by Paolo Virno and by Hardt and Negri is more helpful in drawing off aspects of smartness from their embeddedness within naturalistic and neoliberal assumptions, yet it is not clear to us that these authors successfully engage the ecological dimension of smartness, which is absolutely essential to its current appeal. See Paolo Virno, *A Grammar of the Multitude: For an Analysis of Contemporary Forms of Life* (New York: Semiotext(e), 2004), and Hardt and Negri, *Empire*, as well as *Multitude: War and Democracy in the Age of Empire* (New York: Penguin Press, 2004), and *Commonwealth* (Cambridge, MA: Belknap Press of Harvard University Press, 2009).

12. See especially Foucault, *The History of Sexuality* 1:25–26, 139–147; *Society Must Be Defended*, 239–264; and *Security, Territory, Population*, 38–44, 62–79.

13. Deleuze, "Postscript on the Societies of Control."

14. On financialization and computation, see Michael Lewis, *Flash Boys: A Wall Street Revolt* (New York: W. W. Norton, 2014), and Donald A. MacKenzie, *An Engine, Not a Camera: How Financial Models Shape Markets* (Cambridge, MA: MIT Press, 2006).

15. On the smart home, see Lynn Spigel, "Designing the Smart House: Posthuman Domesticity and Conspicuous Production," in *Electronic Elsewheres: Media, Technology, and the Experience of Social Space*, ed. Chris Berry, Soyoung Kim, and Lynn Spigel (Minneapolis: University of Minnesota Press, 2009), 55–92.

16. There is considerable work—some very critical and some very utopian—on the "smart" city, smart city projects, and "smart" or big data infrastructures. For a sampling, see Rob Kitchin, *The Data Revolution: Big Data, Open Data, Data Infrastructures and Their Consequences* (London: Sage, 2014); Anthony M. Townsend, *Smart Cities: Big Data, Civic Hackers, and the Quest for a New Utopia* (New York: W. W. Norton, 2014); Carlo Ratti and Matthew Claudel, *The City of Tomorrow: Sensors, Networks, Hackers, and the Future of Urban Life* (New Haven, CT: Yale University Press, 2016); Adam Greenfield, *Against the Smart City (the City Is Here for You to Use)* (New York: Do Projects, 2013); Shannon Mattern, *Deep Mapping the Media City* (Minneapolis: University of Minnesota Press, 2015); and Richard Sennett, "The Stupefying Smart City" (paper presented at the Urban Age Electric City Conference, London School of Economics, December 2012).

17. See Keller Easterling, *Extrastatecraft: The Power of Infrastructure Space* (New York: Verso, 2014); and Ned Rossiter, *Software, Infrastructure, Labor: A Media Theory of Logistical Nightmares* (New York: Routledge, 2016).

18. For more on the "logistical" city and free-trade zones, see Brett Nielsen and Ned Rossiter, "The Logistical City," *Transit Labour: Circuits, Regions, Borders*, no. 3 (August 2011): 2–5; Aiwha Ong, "Introduction: Worlding Cities, or the Art of Being Global," in *Worlding Cities, or the Art of Being Global*, ed. Aiwha Ong and Ananya Roy (London: Routledge, 2011), 1–26; Saskia Sassen, *Expulsions: Brutality and Complexity in the Global Economy* (Cambridge, MA: Harvard University Press, 2014); Manuel Castells, *The Rise of the Network Society* (New York: Wiley-Blackwell, 2000); Deborah Cowen, *The Deadly Life of Logistics: Mapping Violence in Global Trade* (Minneapolis: University of Minnesota Press, 2014); and David Harvey, *Spaces of Capital* (London: Routledge, 2012).

19. Orit Halpern, Jesse LeCavalier, and Nerea Calvillo, "Test-Bed Urbanism," *Public Culture* 25, no. 2 (March 2013): 274.

20. Smartness thus partakes in what Shannon Mattern calls methodolatry—namely, a constant obsession with methods and measurement to assess prototypes that are never completed; hence, results are assessed without any clear final metric or end point. See Shannon Mattern, "Methodolatry and the Art of Measure," *Places Journal*, November 2013, https://placesjournal.org/article/methodolatry-and-the-art-of-measure/.

21. See, for example, Anson Rabinbach, *The Human Motor: Energy, Fatigue, and the Origins of Modernity* (Berkeley: University of California Press, 1992).

22. We discuss the history of the term "optimization" in chapter 3.

23. Dan Simon, *Evolutionary Optimization Algorithms: Biologically-Inspired and Population-Based Approaches to Computer Intelligence* (Hoboken, NJ: John Wiley and Sons, 2013), 20–21. Marvin Minsky made similar points in his seminal 1960 article "Steps toward Artificial Intelligence," *Proceedings of the IRE* 49, no. 1 (January 1961): 8–30.

24. In the absence of any way to calculate absolute maxima and minima, the belief that smartness nevertheless locates the "best" solutions can be supported technically by running different optimization algorithms on "benchmark" problems—that is, problems that contain numerous local maxima and minima but for which the absolute maximum or minimum *is* known—in order to determine how well a particular algorithm performs on a given kind of problem. If the algorithm runs well on a benchmark problem, then it is presumed likely to run well on similar real-world problems.

25. David B. Fogel, "An Introduction to Simulated Evolutionary Optimization," *IEEE Transactions of Neural Networks* 5, no. 1 (1994): 3. The issue of *IEEE Transactions of Neural Networks* in which this essay appears, titled *Evolutionary Computing: The Fossil Record*, establishes the importance of Mayr's evolutionary population thinking for this approach to computing (e.g., pp. xi, 1, 11, etc.).

26. Louise Amoore, "Data Derivatives: On the Emergence of a Security Risk Calculus for Our Times," *Theory, Culture and Society* 28, no. 6 (2011): 24–43.

27. Stephanie Wakefield and Bruce Braun, "Living Infrastructure, Government, and Destitute Power" (unpublished paper, Anthropology of the Anthropocene, Concordia University, October 23, 2015), 1–24, 7.

28. D. E. Alexander, "Resilience and Disaster Risk Reduction: An Etymological Journey," *Natural Hazards and Earth System Sciences* 13 (2013): 2707–2716. See also Jeremy Walker and Melinda Cooper, "Genealogies of Resilience: From Systems Ecology to the Political Economy of Crisis Adaptation," *Security Dialogue* 2 (2001): 143–160.

29. C. S. Holling, "Resilience and Stability of Ecological Systems," *Annual Review of Ecological Systems* 4 (1973): 1–23, 21.

30. Holling, "Resilience and Stability of Ecological Systems," 21.

31. Resilience is not equivalent to robustness. As Alexander R. Galloway notes in *Protocol: How Control Exists after Decentralization* (Cambridge, MA: MIT Press, 2004), "robustness" is a defining feature of the technical concept of protocol, which is central

to the computational dimension of smart infrastructures (43–46). However, insofar as robustness refers to the ability of a system to retain its original configuration despite confusing input, it is analogous to what Holling called "stability," rather than resilience. Robustness is thus just one of many technical means for enabling resilient systems.

32. Jennifer Gabrys, *Program Earth: Environmental Sensing Technology and the Making of a Computational Planet* (Minneapolis: University of Minnesota Press, 2016), 4; Benjamin Bratton, *The Stack: On Software and Sovereignty* (Cambridge, MA: MIT Press, 2016), 3–5.

33. For literature on resilience in finance and on economic and development policies, see Melinda Cooper, "Turbulent Worlds: Financial Markets and Environmental Crisis," *Theory, Culture and Society* 27, no. 2–3 (2010): 167–190; and Stephanie Wakefield and Bruce Braun, "Governing the Resilient City," *Environment and Planning D: Society and Space* 32, no. 1 (2014): 4–11.

34. Barry Bergdoll, "Introductory Statement," Museum of Modern Art, accessed November 3, 2016, https://www.moma.org/explore/inside_out/rising-currents?x -iframe=true#description; emphasis added. The original website has now disappeared, but similar materials can be accessed at https://www.moma.org/calendar /exhibitions/1028.

35. Frank Knight, *Risk, Uncertainty, and Profit* (Boston: Houghton Mifflin, 1921).

36. As Joseph Vogl notes in "Taming Time: Media of Financialization," *Grey Room* 46 (2012): 72–83, this seems unlikely to be a successful long-term strategy. Yet the logic of the demo fundamental to resilience ensures that even a massive and widespread financial failure, such as the one that began in 2008, can be treated as simply useful material for subsequent versions of the demo; see Mirowski, *Never Let a Serious Crisis Go to Waste*.

37. Among histories of cybernetics, Philip Mirowski's *Machine Dreams: Economics Becomes a Cyborg Science* (Cambridge: Cambridge University Press, 2002) is especially important for us, as Mirowski stresses a connection between cybernetics and economics absent from many other accounts. Our genealogy also draws in multiple places on Lorraine Daston's multiple histories of scientific reason, rationality, and rules, which range from her earlier *Classical Probability in the Enlightenment* (Princeton, NJ: Princeton University Press, 1995) to her more recent work, *How Reason Almost Lost Its Mind*, coauthored with Erickson, Klein, Lemov, Sturm, and Gordin.

38. Paul N. Edwards, *A Vast Machine: Computer Models, Climate Data, and the Politics of Global Warming* (Cambridge, MA: MIT Press, 2010). Because the smartness mandate is different in kind from climate modeling—the smartness mandate, for example, is not the object of a specific scientific discipline—our method necessarily differs from Edwards. At the same time, his account resonates with ours at multiple points, especially in connection with the ecological concerns and data-gathering techniques that emerged in the 1960s and eventually became part of the smartness mandate.

39. See for example Peder Anker, *Imperial Ecology: Environmental Order in the British Empire, 1895–1945* (Cambridge, MA: Harvard University Press, 2001); Anker, *From*

Bauhaus to Ecohouse: A History of Ecological Design (Baton Rouge: Louisiana State University Press, 2010); and Etienne Benson, *Surroundings: A History of Environments and Environmentalisms* (Chicago: University of Chicago Press, 2020).

40. Fred Turner, *The Democratic Surround: Multimedia and American Liberalism from World War II to the Psychedelic Sixties* (Chicago: University of Chicago Press, 2013).

41. See Yuriko Furuhata, *Climatic Media: Transpacific Experiments in Atmospheric Control* (Durham, NC: Duke University Press, 2022); Nicole Starosielski, *Media Hot and Cold* (Durham, NC: Duke University Press, 2021); Daniel Barber, *Modern Architecture and Climate: Design before Air Conditioning* (Princeton, NJ: Princeton University Press, 2020).

42. Lisa Parks and Caren Kaplan, introduction to *Life in the Age of Drone Warfare*, ed. Lisa Parks and Caren Kaplan (Durham, NC: Duke University Press, 2017), 10.

43. Shannon Mattern, *Code and Clay, Data and Dirt: Five Thousand Years of Urban Media* (Minneapolis: University of Minnesota Press, 2017), xvi.

44. Donella H Meadows, Dennis L. Meadows, Jørgen Randers, and William W. Behrens, *The Limits to Growth: A Report for the Club of Rome's Project on the Predicament of Mankind* (New York: Universe Books, 1972).

CHAPTER 1

1. Naomi Klein, "Screen New Deal," *The Intercept*, May 8, 2020, https://theintercept .com/2020/05/08/andrew-cuomo-eric-schmidt-coronavirus-tech-shock-doctrine/.

2. For histories of shock as a concept in psychology, see Marc-Antoine Crocq and Louis Crocq, "From Shell Shock and War Neurosis to Posttraumatic Stress Disorder: A History of Psychotraumatology," *Dialogues in Clinical Neuroscience* 2, no. 1 (2000): 47–55; Mark S. Micale and Paul Lerner, eds., *Traumatic Pasts: History, Psychiatry, and Trauma in the Modern Age, 1870–1930* (Cambridge: Cambridge University Press, 2010).

3. Francis Bacon, *The Essays, or Councils, Civil and Moral, of Sir Francis Bacon, Lord Verulam, Viscount St. Alban with a Table of the Colours of Good and Evil, and a Discourse of the Wisdom of the Ancients: To This Edition Is Added the Character of Queen Elizabeth, Never before Printed in English* (London: Printed for H. Herringman, R. Scot, R. Chiswell, A. Swalle, and R. Bentley, 1696), 38.

4. Thomas Malthus, *An Essay on the Principle of Population, as It Affects the Future Improvement of Society. With Remarks on the Speculations of Mr. Godwin, M. Condorcet, and Other Writers* (London: J. Johnson, 1798), 18.

5. Malthus, *An Essay on the Principle of Population*, 14.

6. In later editions of his text, Malthus grudgingly acknowledged, via the introduction of the concept of *moral restraint*, that at least some members of populations could work against these drives. However, he did not see this as a possibility open to enough members of a population to make a difference. On the question of what changes in practice Malthus thought knowledge about population dynamics could produce, see Robert Mitchell, "Regulating Life: Romanticism, Science, and the Liberal Imagination," *European Romantic Review* 29, no. 3 (2018): 275–293.

7. See, for example, Piers J. Hale, *Political Descent: Malthus, Mutualism, and the Politics of Evolution in Victorian England* (Chicago: University of Chicago Press, 2014).

8. On the role of Lotka and Volterra in ecosystem ecology, see Joel B. Hagen, *An Entangled Bank: The Origins of Ecosystem Ecology* (New Brunswick, NJ: Rutgers University Press, 1992), 70–72, 125–136; see also Philip Kreager, "Darwin and Lotka: Two Concepts of Population," *Demographic Research* 21, no. 2 (2009): 469–502.

9. On social Darwinism, see Richard Hofstadter, *Social Darwinism in American Thought* (New York: G. Braziller, 1959); Mike Hawkins, *Social Darwinism in European and American Thought, 1860–1945: Nature as Model and Nature as Threat* (Cambridge: Cambridge University Press, 1997).

10. As Michel Foucault established, this second model of population was essential to the development of biopolitics, which focused on gathering data about population-level norms and regularities with the goal of then altering those norms and regularities. One of Foucault's key examples was eighteenth-century debates about smallpox inoculation, which in some cases involved parsing populations into different groups (for example, based on age) and determining the efficacy of inoculation for each subgroup. See Foucault, *Security, Territory, Population: Lectures at the Collège de France, 1977–78*, ed. Michel Senellart, trans. Graham Burchnell (New York: Palgrave Macmillan, 2007), 55–86.

11. See Lorraine Daston, *Classical Probability in the Enlightenment* (Princeton, NJ: Princeton University Press, 1995).

12. On the rise of probabilistic thinking and techniques in the nineteenth century, see Daston, *Classical Probability in the Enlightenment*; Theodore M. Porter, *The Rise of Statistical Thinking, 1820–1900* (Princeton, NJ: Princeton University Press, 1986); Ian Hacking, *The Taming of Chance* (Cambridge: Cambridge University Press, 1990); Alain Desrosières, *The Politics of Large Numbers: A History of Statistical Reasoning*, trans. Camille Naish (Cambridge, MA: Harvard University Press, 1998). On the development of social insurance structures within the French welfare state, see François Ewald, *The Birth of Solidarity: The History of the French Welfare State* (Durham, NC: Duke University Press, 2020). On the development of workers' compensation in the US, see Paul B. Bellamy, *A History of Workmen's Compensation, 1898–1915: From Courtroom to Boardroom* (New York: Garland Publishing, 1997); John Fabian Witt, *The Accidental Republic: Crippled Workingmen, Destitute Widows, and the Remaking of American Law* (Cambridge, MA: Harvard University Press, 2004). See also Sharon Bertsch McGrayne, *The Theory That Would Not Die: How Bayes' Rule Cracked the Enigma Code, Hunted Down Russian Submarines, and Emerged Triumphant from Two Centuries of Controversy* (New Haven, CT: Yale University Press, 2011), who stresses that in the US the swift passage of workers' compensation legislation and the consequent lack of prior data on which to base insurance premiums encouraged a turn to Bayesian techniques for setting premiums. This is an important point, for it underscores the extent to which actuarial approaches to population also served as vectors for a certain form of mathematical modeling.

13. For an extended discussion of this aspect of biobanks, see Robert Mitchell and Catherine Waldby, "National Biobanks: Clinical Labour, Risk Production, and the

Creation of Biovalue," *Science, Technology, and Human Values* 35, no. 3 (2010): 330–355; and Robert Mitchell, "U.S. Biobanking Strategies and Biomedical Immaterial Labor," *BioSocieties* 7, no. 3 (2012): 224–244.

14. On ways in which research hospital clinics are repurposed to meet research demands, see Mitchell, "U.S. Biobanking Strategies."

15. The concept of *risk factors* emerged in the mid-twentieth century in the context of the population-oriented Framingham Study; see W. G. Rothstein, *Public Health and the Risk Factor: A History of an Uneven Medical Revolution* (Rochester, NY: University of Rochester Press, 2003).

16. As Joseph Dumit documents in *Drugs for Life: How Pharmaceutical Companies Define Our Health* (Durham, NC: Duke University Press, 2012), some physicians hope that eventually all members of the population will take multiple drugs "for life" in order to reduce health risks.

17. Charles Darwin, *The Origin of Species by Means of Natural Selection; or the Preservation of Favoured Races in the Struggle for Life* (London: John Murray, 1859), 5. In *Political Descent*, Hale provides a helpful reading of the role of Malthus in Darwin's theory and the reception of the latter by late nineteenth- and early twentieth-century biological and social theorists.

18. Ernst Mayr, *Systematics and the Origin of Species: From the Viewpoint of a Zoologist* (New York: Columbia University Press, 1942), 5.

19. Mayr, *Systematics and the Origin of Species*, 8.

20. Mayr, *Systematics and the Origin of Species*, 6–8.

21. Ernst Mayr, "Darwin and the Evolutionary Theory in Biology," in *Evolution and Anthropology: A Centennial Appraisal*, ed. B. J. Meggers (Washington, DC: Anthropological Society of Washington), 2.

22. Mayr, *Systematics and the Origin of Species*, 99.

23. For Mayr's use of Wright's concept of the fitness landscape, see Mayr, *Systematics and the Origin of Species*, 99.

24. In his hugely influential article "Steps toward Artificial Intelligence," *Proceedings of the IRE* 49, no. 1 (1961): 8–30, computer scientist Marvin Minsky discussed both what he called the "hill-climbing" method and problems with that method (9–10) and referred back to both W. R. Ashby and Norbert Wiener's work in the 1940s and 1950s as key theorists for this approach. Minsky's article was one of many in this period—a number of which appeared in various Institute of Radio Engineers (IRE) journals—that provided surveys of what were often called adaptive or self-optimization methods, which mimicked the adjustment and learning capacities of living beings. See, e.g., J. A. Aseltine et al., "A Survey of Adaptive Control Systems," *IRE Transactions on Automatic Control* 6, no. 1 (1958): 102–108. A Google Books search for "phase space" makes clear that both Ashby and Sewall's use of that concept originated in late nineteenth- and early twentieth-century "statistical" (i.e., quantum) physics.

25. For an example of the former, see J. L. Crosby, "Computers in the Study of Evolution," *Science Progress* 55 (1967): 279–292; for an example of the latter, see L. I.

Fogel, A. J. Owens, and M. I. Walsh, "Artificial Intelligence through a Simulation of Evolution," in *Biophysics and Cybernetic Systems: Proceedings of the 2nd Cybernetic Sciences Symposium*, ed. M. Maxfield, A. Callahan, and L. J. Fogel (Washington, DC: Spartan Books, 1965), 131–155, reprinted in Fogel, *Evolutionary Computation*, 230–254. Wright's theory of fitness landscapes has subsequently been extended, primarily through the intermediary of Stuart A. Kaufmann's work, especially *The Origins of Order: Self-Organization and Selection in Evolution* (Oxford: Oxford University Press, 1993) to many other fields, including economics, organization and management sciences, psychology, and political science. For a survey of the influence of the concept of fitness landscapes, see Lasse Gerrits and Peter Marks, "The Evolution of Wright's (1932) Adaptive Field to Contemporary Interpretations and Uses of Fitness Landscapes in the Social Sciences," *Biology and Philosophy* 30 (2015): 459–479.

26. For the context in which Hayek composed and published *The Road to Serfdom*, see Bruce Caldwell, introduction to F. A. Hayek, *The Road to Serfdom: Text and Documents*, ed. Bruce Caldwell (Chicago: University of Chicago Press, 2007), 1–33.

27. On the history of the Mont Pelerin Society, see Philip Mirowski and Dieter Plehwe, eds., *The Road from Mont Pèlerin: The Making of the Neoliberal Thought Collective* (Cambridge, MA: Harvard University Press, 2009).

28. Friedrich Hayek, "The Use of Knowledge in Society," *American Economic Review* 35, no. 4 (1945): 519.

29. Hayek, "The Use of Knowledge in Society," 521–522.

30. Hayek, "The Use of Knowledge in Society," 521–522.

31. Hayek, "The Use of Knowledge in Society," 523, 526.

32. Hayek, "The Use of Knowledge in Society," 521. That is, competition is the most "efficient" means of planning since it makes the "full[est] use of the existing knowledge" (521).

33. Hayek, "The Use of Knowledge in Society," 525.

34. Hayek, "The Use of Knowledge in Society," 526.

35. Philip Mirowski, "Twelve Theses Concerning the History of Postwar Neoclassical Price Theory," *History of Political Economy* 38 (2006): 343–379.

36. On the multiple origins of evolutionary computing, see David B. Fogel, *Evolutionary Computation: The Fossil Record* (Hoboken, NJ: John Wiley & Sons, 1998), xi–xii. Fogel dates the term "evolutionary computation" to 1991 (1).

37. For early attempts to use computers to model natural evolution, see A. S. Fraser, "Simulation of Genetic Systems by Automatic Digital Systems. I. Introduction," *Australian Journal of Biological Sciences* 10 (1957): 484–491, reprinted in Fogel, *Evolutionary Computation*, 87–94; and Crosby, "Computers in the Study of Evolution," reprinted in Fogel, *Evolutionary Computation*, 95–108. For early attempts to "evolve" industrial productivity, see G. E. P. Box, "Evolutionary Operation: A Method for Increasing Industrial Productivity," *Applied Statistics* 6, no. 2 (1957): 81–101, reprinted in Fogel, *Evolutionary Computation*, 121–141. For early attempts to create artificial intelligence, see Fogel, Owens, and Walsh, "Artificial Intelligence through a Simulation

of Evolution," 131–155, reprinted in Fogel, *Evolutionary Computation*, 230–254. On fluid mechanics, see I. Rechenberg, "Cybernetic Solution Path of an Experimental Problem," *Royal Aircraft Establishment, Library Translations*, no. 1122 (1965), reprinted in Fogel, *Evolutionary Computation*, 297–309.

38. The importance of cybernetics was stressed explicitly in some of these early articles: Rechenberg, for example, contended that "in cybernetics it is axiomatic that common theories can be applied to apparently widely separated fields of science. The increasingly evident points in common between biology as the theory of organisms and technology as the theory of mechanisms provide a good example of this" (Rechenberg, "Cybernetic Solution Path," reprinted in Fogel, *Evolutionary Computation*, 301). And though computer scientists and biologists did not always directly influence or work with one another in this early period of evolutionary computation, there were occasions in which they presented their work to one another. A notable example was a 1966 conference titled "Mathematical Challenges to the Neo-Darwinian Interpretation of Evolution," held at the Wistar Institute of Anatomy and Biology. This conference brought together biologists such as Richard Lewontin, Ernst Mayr, Peter Medawar, Sewall Wright, and C. H. Waddington with computer scientists and bioengineers such as Murray Eden, W. Bossert, and Nils Barracelli. See Paul S. Moorhead and Martin M. Kaplan, eds., *Mathematical Challenges to the Neo-Darwinian Interpretation of Evolution* (Philadelphia: Wistar Institute Press, 1967).

39. Fogel, *Evolutionary Computation*, 3.

40. Fogel, *Evolutionary Computation*, 4.

41. This example is taken from Dan Simon, *Evolutionary Optimization Algorithms: Biologically-Inspired and Population-Based Approaches to Computer Intelligence* (Hoboken, NJ: John Wiley & Sons, 2013), 44–49.

42. For an early example of this approach to industrial productivity, see Box, "Evolutionary Operation."

43. As Frank Rosenblatt noted near the start of his first article on perceptrons, "the most suggestive work, for the standpoint of the following theory [i.e., the theory that Rosenblatt develops in the article], is that of Hebb and Hayek." See Rosenblatt, "The Perceptron: A Probabilistic Model for Information Storage and Organization in the Brain," *Psychological Review* 65, no. 6 (1958): 386–408, at 388.

44. Rosenblatt, "Perceptron," 388–389.

45. Rosenblatt, "Perceptron," 388.

46. Rosenblatt discusses the role of training in "The Perceptron" (395) and at more length in *Principles of Neurodynamics: Perceptrons and the Theory of Brain Mechanisms* (Washington, DC: Spartan Books, 1962).

47. Rosenblatt, *Principles of Neurodynamics*, 388.

48. Hayek, "The Use of Knowledge in Society," 526.

49. Hayek, "The Use of Knowledge in Society," 522.

50. Hayek, "The Use of Knowledge in Society," 526.

51. Rosenblatt, *Principles of Neurodynamics*, 17.

52. Rosenblatt, *Principles of Neurodynamics*, 19–20; emphasis added.

53. When studying perceptrons, the "object of analysis is an experimental system which includes the perceptron, a defined environment, and a training procedure or agency" (Rosenblatt, *Principles of Neurodynamics*, 27–28).

54. For discussion of these data sets, see Aurélien Géron, *Hands-On Machine Learning with Scikit-Learn and TensorFlow: Concepts, Tools, and Techniques to Build Intelligent Systems* (Sebastopol, CA: O'Reilly Media, 2017), 33–34.

55. Donella H. Meadows et al., *The Limits to Growth: A Report for the Club of Rome's Project on the Predicament of Mankind* (New York: Universe Books, 1972), 9.

56. Meadows et al., *The Limits to Growth*, 9.

57. Meadows et al., *The Limits to Growth*, 12.

58. For an early discussion of the background of, as well as responses to, *The Limits to Growth*, see Francis Sandbach, "The Rise and Fall of the Limits to Growth Debate," *Social Studies of Science* 8, no. 4 (1978): 495–520; for a more recent discussion, see Elodie Vieille Blanchard, "Modelling the Future: An Overview of the 'Limits to Growth' Debate," *Centaurus* 52 (2010): 91–116. For a discussion focused primarily on Forrester, see Fernando Elichirigoity, *Planet Management: Limits to Growth, Computer Simulation, and the Emergence of Global Spaces* (Evanston, IL: Northwestern University Press, 1999). See also Melinda Cooper, *Life as Surplus: Biotechnology and Capitalism in the Neoliberal Era* (Seattle: University of Washington Press, 2008), 15–50. As Elichirigoity notes in *Planet Management*, Malthus's approach to population was the subject of some of the early presentations at MIT that led to *The Limits to Growth* (88).

59. Jay W. Forrester, *Industrial Dynamics* (Cambridge, MA: MIT Press, 1961).

60. Meadows et al., *The Limits to Growth*, 23.

61. Meadows et al., *The Limits to Growth*, 24.

62. On President Carter's commission, see Cooper, *Life as Surplus*, 17.

63. For an overview of these various responses, see Blanchard, "Modelling the Future."

64. On the connection between *The Limits to Growth* and the development of the Chinese one-child policy, see Susan Greenhalgh and Edwin A. Winckler, *Governing China's Population: From Leninist to Neoliberal Biopolitics* (Stanford, CA: Stanford University Press, 2005), 291–300. See also Carole R. McCann, *Figuring the Population Bomb: Gender and Demography in the Mid-twentieth Century* (Seattle: University of Washington Press, 2016).

65. For the linkages between the world dynamics computer modeling employed in *The Limits to Growth* and climate modeling, see Paul N. Edwards, *A Vast Machine: Computer Models, Climate Data, and the Politics of Global Warming* (Cambridge, MA: MIT Press, 2010), 366–372.

66. On neoliberal criticisms of *The Limits to Growth*, see Cooper, *Life as Surplus*, 15–50.

67. The original academic papers that described Google's system include Sergey Brin and Lawrence Page, "The Anatomy of a Large-Scale Hypertextual Web Search Engine" (working paper, Stanford InfoLab, Stanford University, accessed March 6, 2016,

http://infolab.stanford.edu/~backrub/google.html); and Lawrence Page, Sergey Brin, Rajeev Motwani, and Terry Winograd, "The PageRank Citation Ranking: Bringing Order to the Web," Technical Report, Stanford InfoLab, 1999, http://ilpubs.stanford .edu:8090/422/. For accounts of the history and technical aspects of Google's Page-Rank approach, see Amy N. Lanville and Carl D. Meyer, *Google's PageRank and Beyond: The Science of Search Engine Rankings* (Princeton, NJ: Princeton University Press, 2006); Michael W. Berry and Murray Browne, *Understanding Search Engines: Mathematical Modeling and Text Retrieval* (Philadelphia: Society for Industrial and Applied Mathematics, 2005). For a useful critical account of the value logic of PageRank, see Matteo Pasquinelli, "Google's PageRank Algorithm: A Diagram of the Cognitive Capitalism and the Rentier of the Common Intellect," in *Deep Search: The Politics of Search Beyond Google*, ed. Konrad Becker and Felix Stalder (London: Transaction Publishers, 2009), 152–162, http://matteopasquinelli.com/google-pagerank-algorithm/.

68. It is worth stressing that in academia the formal system of numerical article and journal ranking dates back only to the mid-1960s and has developed in a decidedly neoliberal direction in the last two decades; see Philip Mirowski, *Science-Mart: Privatizing American Science* (Cambridge, MA: Harvard University Press, 2011), 266–287. As a consequence, PageRank was based on a very specific, recent—and, as Mirowski convincingly argues—neoliberal model of academic expert communities.

69. "One can simply think of every link as like an academic citation," and thus a "major page like http://www.yahoo.com/ will have tens of thousands of backlinks (or citations) pointing to it": Page et al., "The PageRank Citation Ranking," 2.

70. Page et al., "The PageRank Citation Ranking," 1. PageRank was thus a "method for computing a ranking for every web page based on the graph of the web" (2) with "graph" here understood to mean the number and nature of links to every page.

71. Page et al., "The PageRank Citation Ranking," 2.

72. Nicholas Carr, *The Big Switch: Rewiring the World, from Edison to Google* (New York: W. W. Norton, 2008).

73. Pasquinelli, "Google's PageRank Algorithm."

74. Page et al., "The PageRank Citation Ranking," 12.

75. Chicago school neoliberal Milton Friedman stressed this point in his classic essay "The Methodology of Positive Economics," included in Friedman, *Essays in Positive Economics* (Chicago: University of Chicago Press, 1953), 3–43.

76. On the emergence of Wikipedia, see Nathaniel Tkacz, *Wikipedia and the Politics of Openness* (Chicago: University of Chicago Press, 2014). (Tracz emphasizes that Wikipedia redefined, rather than eliminated, expertise—a point that holds more generally for both Hayek, for whom the neoliberal economist's expertise must be trusted when it comes to clarifying how to optimize the market, and for the smart developments we discuss.) For a critical account of the development of open-source and citizen science that emphasizes the ways in which these efforts have been encouraged and coordinated with for-profit platforms, see Philip Mirowski, "The Future(s) of Open Science," *Social Studies of Science* 48, no. 2 (2018): 171–203.

77. Klein, "Screen New Deal."

CHAPTER 2

1. From the start, the competition was plagued by criticism: the cost of entry for junior architects or designers was prohibitive, and all the chosen designs came from well-known and well-established architectural firms linked to the most prestigious and storied architecture schools in New York, such as Columbia University. In addition, the projected designs were not evaluated for feasibility or environmental impact and were critiqued for supporting progrowth ideologies and the dispossession of urban poor and minority communities. See Justin Davidson, "MoMA's Schemes for Fixing Urban Problems Are Either Too Dainty or Too Sweeping," *The Vulture*, November 20, 2014, https://www.vulture.com/2014/11/review-momas-uneven-growth.html; Stephanie Wakefield and Bruce Braun, "Oystertecture: Infrastructure, Profanation, and the Sacred Figure of the Human," in *Infrastructure, Environment and Life in the Anthropocene*, ed. Kregg Hetherington (Durham, NC: Duke University Press, 2019), 193–215.

2. Barry Bergdoll, "Introductory Statement," Museum of Modern Art, accessed November 3, 2016, https://www.moma.org/explore/inside_out/rising-currents?x-iframe=true #description. The website no longer exists, but similar materials can be accessed at https://www.moma.org/calendar/exhibitions/1028.

3. Kate Orff, "Oyster-tecture," Scape Landscape Architecture DCP, accessed December 10, 2020, https://www.scapestudio.com/projects/oyster-tecture/.

4. nArchitects, "Aqueous City," Museum of Modern Art, accessed April 16, 2017, http://narchitects.com/work/moma-rising-currents/. The website no longer exists, but the project is available (as of 2021) at https://www.youtube.com/watch?v=O7folAgxX -gaqueous.

5. Lisa Wirthman, "NYC's Hudson Yards Looks towards the Future of Smart Development," *Forbes*, October 3, 2018.

6. Aria Bendix, "New York's $25 Billion Megadevelopment Can Withstand a Superstorm or Terrorist Attack—Even If the Entire City Shuts Down," Business Insider, March 14, 2019, https://www.businessinsider.com/hudson-yards-can-withstand -superstorm-terrorist-attack-2019-3; Matthew Schuerman, "Climate Change Fears Meet Development at the New Hudson Yards," WNYC News, December 12, 2012, https://www.wnyc.org/story/256763-underside-developing-hudson-yards/.

7. See also Orit Halpern and Gokce Günel's discussion of "preemptive hopefulness" in their "Demoing unto Death: Smart Cities, Environment, and Preemptive Hope," *FibreCulture*, no. 29 (2017), http://twentynine.fibreculturejournal.org/fcj-215 -demoing-unto-death-smart-cities-environment-and-preemptive-hope/.

8. Ash Amin, "On Urban Failure," *Social Research* 83, no. 3 (2016): 778. See also Ayona Datta, "Postcolonial Urban Futures: Imagining and Governing India's Smart Urban Age," *Environment and Planning D: Society and Space* 37, no. 3 (2019): 393–410.

9. As we hope will be evident in what follows, we treat Negroponte in this chapter primarily as a channel or conduit for interests and tendencies (and tensions) already present at MIT, in Boston, and in 1960s and 1970s urban planning, architecture, and computing more generally. Many of the concepts that we attribute specifically to

Negroponte can also be found in the work of other MIT faculty. For example, Warren Brodey, originally a psychiatrist and psychoanalyst and then an MIT cybernetics enthusiast before eventually becoming a Maoist organizer abroad, also promoted notions of "soft" architecture and human-machine urban design in the late 1960s. He was specifically acknowledged by Negroponte in *The Architecture Machine* and cited several times in both *The Architecture Machine* and *Soft Architecture Machines*.

10. Molly Wright Steenson, "Architectures of Information: Christopher Alexander, Cedric Price, and Nicholas Negroponte and the MIT Architecture Machine Group" (PhD diss., Princeton University, 2014). See also Molly Wright Steenson, *Architectural Intelligence* (Cambridge, MA: MIT Press, 2017).

11. Nicholas Negroponte, *The Architecture Machine* (Cambridge, MA: MIT Press, 1970), i.

12. Negroponte, *The Architecture Machine*.

13. Negroponte, *The Architecture Machine*, 55.

14. Negroponte, *The Architecture Machine*.

15. Negroponte, *The Architecture Machine*, 57.

16. Significantly, this demo project involved no actual computers, for at the time, computers could not handle such sophisticated questions, and the test was in fact secretly run by a human being. The participants, however, were kept ignorant of this fact, believing that they were speaking to a machine. One can read, therefore, the whole test as an interface to and a demo of what a computationally aided interaction in service of a responsive government might be. Negroponte, *The Architecture Machine*, 55–58.

17. Negroponte, *The Architecture Machine*, 55–56.

18. Negroponte, *The Architecture Machine*, 56.

19. Nicholas Negroponte, "Systems of Urban Growth" (BA thesis, Massachusetts Institute of Technology, 1965). In the preface to his thesis, Negroponte noted that the topic of "population"—which for him meant the "study of how populations live, what populations want, and, primarily, how populations expand"—provided him with a "vantage point" that allowed him to develop his more primary interest in "the city as an organism," i.e., "a machine of communication."

20. Stephen Graham and Simon Marvin, *Splintering Urbanism: Networked Infrastructures, Technological Mobilities and the Urban Condition* (London: Routledge, 2001), 111.

21. See Peter Dreier, "Downtown Development and Urban Reform: The Politics of Boston's Linkage Policy," *Urban Affairs Quarterly* 26, no. 3 (1991): 354–370; Melvin King, *Chain of Change: Struggles for Black Community Development* (Boston: South End Press, 1981). See also "MIT's Community Fellows Program," MIT School of Architecture and Planning, accessed December 10, 2015, http://web.mit.edu/mhking/www/cfp.html.

22. Edward Murray Bassett, *Zoning* (1922; repr., New York: Russell Sage Foundation, 1940), 330.

23. Frank Backus Williams, *Building Regulation by Districts: The Lesson of Berlin* (New York: National Housing Association, 1914), 1.

24. Negroponte, *The Architecture Machine*, 67.

25. Nicholas Negroponte, *Soft Architecture Machines* (Cambridge, MA: MIT Press, 1975), 103.

26. Negroponte, *Soft Architecture Machines*, 119.

27. Jennifer Light, *From Warfare to Welfare: Defense Intellectuals and Urban Problems in Cold War America* (Baltimore: John Hopkins University Press, 2003).

28. Orit Halpern, *Beautiful Data: A History of Vision and Reason since 1945* (Durham, NC: Duke University Press, 2015), 79–199; Jay W. Forrester, *Industrial Dynamics* (Cambridge, MA: MIT Press, 1961); Jay W. Forrester, *Urban Dynamics* (Cambridge, MA: MIT Press, 1969); Kevin Lynch, *The Image of the City* (Cambridge, MA: MIT Press, 1960).

29. Matthew W. Hughey, "Of Riots and Racism: Fifty Years since the Best Laid Schemes of the Kerner Commission (1968–2018)," *Sociological Forum* 33, no. 3 (2018): 627.

30. Fenwick McKelvey, "The Other Cambridge Analytics: Early 'Artificial Intelligence' in American Political Science," in *The Cultural Life of Machine Learning: An Incursion into Critical AI Studies*, ed. Jonathan Roberge and Michael Castelle (Cham, Switzerland: Palgrave Macmillan/Springer Nature Switzerland AG, 2020), internal quotes from MIT Archives, 132. For more material on Simulmatics and the question of automating democracy, see Jill Lepore, *If Then: How the Simulmatics Corporation Invented the Future* (New York: Liveright, 2020).

31. Forrester, *Urban Dynamics*, 1.

32. Forrester, *Urban Dynamics*, ix–x, 14–18, 107–115.

33. Forrester, *Urban Dynamics*, ix–x, 12–17.

34. Daniel M. Abramson, *Obsolescence: An Architectural History* (Chicago: University of Chicago Press, 2016).

35. Robert Fishman, *Urban Utopias in the Twentieth Century: Ebenezer Howard, Frank Lloyd Wright, and Le Corbusier* (Cambridge, MA: MIT Press, 1982).

36. Shannon Mattern, "Interfacing Urban Intelligence," *Places Journal*, April 14, 2014, https://placesjournal.org/article/interfacing-urban-intelligence/.

37. Negroponte, *The Architecture Machine*, 2.

38. Steve Heims, *The Cybernetics Group* (Cambridge, MA: MIT Press, 1991); Peter Galison, "The Ontology of the Enemy: Norbert Wiener and the Cybernetic Vision," *Critical Inquiry* 21 (1994): 228–266.

39. Norbert Wiener, *Cybernetics; or, Control and Communication in the Animal and the Machine* (Cambridge, MA: MIT Press, 1961).

40. The pair would go to MIT in 1952 at Wiener's behest.

41. Lily E. Kay, "From Logical Neurons to Poetic Embodiments of Mind: Warren McCulloch's Project in Neuroscience," *Science in Context* 14, no. 4 (2001): 591–614.

42. Warren McCulloch and Walter Pitts, "A Logical Calculus of Ideas Immanent in Nervous Activity," in *Embodiments of Mind*, ed. Warren McCulloch (1943; repr., Cambridge, MA: MIT Press, 1970).

43. Frank Rosenblatt, "The Perceptron: A Probabilistic Model for Information Storage and Organization in the Brain," *Psychological Review* 65, no. 6 (1958): 386.

44. Rosenblatt, "The Perceptron," 387.

45. Rosenblatt, "The Perceptron," 388.

46. Oliver G. Selfridge, "Pandemonium: A Paradigm for Learning," in *Proceedings of the Symposium on Mechanisation of Thought Processes*, ed. D. V. Blake and A. M. Uttley (London: Her Majesty's Stationery Office, 1959), 511–529. Arch Mac also engaged the ideas of Marvin Minsky, Warren McCulloch, and Ross Ashby, according to Nicholas Negroponte's accounts at the start of *The Architecture Machine* and *Soft Architecture Machines*.

47. Branden Hookway, *Pandemonium: The Rise of Predatory Locales in the Postwar World* (Princeton, NJ: Princeton Architectural Press, 1999).

48. Selfridge, "Pandemonium," 513.

49. Hookway, *Pandemonium*.

50. Negroponte, *Soft Architecture Machines*, viii.

51. Lian Chikako Chang, "Live Blog—Mohsen Mostafavi in Conversation with Nicholas Negroponte," *Archinect* (blog), January 15, 2013, https://archinect.com/lian/live-blog-mohsen-mostafavi-in-conversation-with-nicholas-negroponte.

52. Andrew Lippman, interview by Orit Halpern, November 25, 2014, MIT Media Lab. The background of the movie map in relation to DARPA is also discussed by Michael Naimark, "Aspen the Verb: Musings on Heritage and Virtuality," *Presence Journal* 15, no. 3 (2006), http://www.naimark.net/writing/aspen.html.

53. Noah Wardrip-Fruin and Nick Montfort, eds., *The New Media Reader* (Cambridge, MA: MIT Press, 2003), chaps. 17, 23; Marisa Rennee Brandt, "War, Trauma, and Technologies of the Self: The Making of Virtual Reality Exposure Therapy" (PhD diss., University of San Diego, 2013).

54. Naimark, "Aspen the Verb"; Michael Naimark, interview by Orit Halpern, August 12, 2013, Cambridge, MA.

55. Naimark, interview. Naimark informed Halpern that they went out at the same time every day to record footage.

56. Lian Chikako Chang, "Live Blog—Mohsen Mostafavi in Conversation with Nicholas Negroponte."

57. Felicity Scott, "Aspen Proving Grounds" (paper presented at Technics and Art: Architecture, Cities, and History after Mumford, Columbia University, 2012).

58. Tristan d'Estrée Sterk, "Building upon Negroponte: A Hybridized Model of Control Suitable for Responsive Architecture," *Automation in Construction* 14, no. 2 (2014): 225–232.

59. Larry D. Busbea, *The Responsive Environment: Design Aesthetics and the Human in the 1970s* (Minneapolis: University of Minnesota Press, 2019), xx.

60. Negroponte, *Soft Architecture Machines*, 103.

61. Negroponte, *Soft Architecture Machines*, 103–104.

62. Rebecca Slayton, "Efficient, Secure, Green: Digital Utopianism and the Challenge of Making the Electrical Grid 'Smart,'" *Information and Culture* 48, no. 4 (2013): 448–478, 461–462.

63. Slayton, "Efficient, Secure, Green," 454.

64. On the role played by *The Limits to Growth*, see Slayton, "Efficient, Secure, Green," 455, 474n43.

65. Slayton, "Efficient, Secure, Green," 457.

66. Slayton, "Efficient, Secure, Green."

67. Slayton, "Efficient, Secure, Green," 458.

68. Kurt E. Yeager, "Creating the Second Electrical Century," *Public Utilities Fortnightly* 126 (1990): 1, cited in Slayton, "Efficient, Secure, Green," 458.

69. Khoi Vu, Miroslav Begovic, and Damir Novosel, "Grids Get Smart Protection and Control," *IEEE Computer Applications in Power* 10, no. 4 (1997): 40, cited in Slayton, "Efficient, Secure, Green," 462.

70. Vu, Begovic, and Novosel, "Grids Get Smart Protection and Control," 41. Slayton notes that after the terrorist attacks on the World Trade Center in 2011, smartness was also charged with preventing terrorist attacks on the electrical grid (Slayton, "Efficient, Secure, Green," 464).

71. As Slayton notes in "Efficient, Secure, Green," "Unfortunately, by [using smart technologies to] operat[e] large areas closer to margins of error, utilities also risked failures on an even larger scale, and blackouts increased in frequency and size through the 1990s and early new millennium" (462).

72. H. Res. 109 (2019), at 8.

73. Winona LaDuke and Deborah Cowen, "Beyond Wiindigo Infrastructure," *South Atlantic Quarterly* 119, no. 2 (2020): 243–268, 245.

74. Dave Lovekin, "Unlocking Clean Energy Opportunities for Indigenous Communities: Federal Funding Will Help Communities Develop Renewable Energy Projects and Transition Off Diesel," Pembina Institute, February 24, 2017, https://www.pembina.org/blog/unlocking-clean-energy-opportunities-indigenous-communities.

75. Krista Allen, "NTUA, Kayenta Solar Project Chart Path to the Future," *Navajo Times*, October 3, 2019, https://navajotimes.com/biz/ntua-kayenta-solar-project-chart-path-to-the-future/.

76. Jannine Anderson, "Navajo Solar Project Sending Power to the Grid in Kayenta, Ariz.," *American Public Power Association*, June 20, 2017; Bonnie Riva Ras, "The Navajo Nation's First Solar Project Is Scaling Up," *Goodnet: Gateway to Doing Good*, January 20, 2019, https://www.goodnet.org/articles/navajo-nations-first-solar-project-scaling-up. LaDuke and Cowen note that the Navajo Nation has seen very few benefits from the agreements signed in the 1920s that promised royalties for uranium, oil, gas, and coal extraction on their lands. Although the Nation provided enough energy to power the entire state of New Mexico 32 times over, 85 percent of Nation households had no electricity before the panels went online in 2017 (256).

77. Allen, "NTUA, Kayenta Solar Project Chart Path to the Future." See also Patty Garcia-Likens, "NTUA, SRP and Navajo Nation Leaders Celebrate Groundbreaking Ceremony for Kayenta II," SRP, August 27, 2018, https://media.srpnet.com/kayenta-ii-groundbreaking-ceremony/.

78. Allen, "NTUA, Kayenta Solar Project Chart Path to the Future."

79. Negroponte, *The Architecture Machine*, 1.

80. See Robert E. Park, "The City as a Social Laboratory," in *Chicago: An Experiment in Social Science Research*, ed. Thomas Vernor Smith and Leonard Dupee White (Chicago: University of Chicago Press, 1929), 1–19, as well as the other essays in that volume. This early twentieth-century University of Chicago sociology approach to experimentation has been an important reference point for several recent geographers who advocate for experimental geography; see, e.g., Matthias Gross and Wolfgang Krohn, "Society as Experiment: Sociological Foundations for a Self-Experimental Society," *History of the Human Sciences* 18, no. 2 (2005): 63–86; and Andrew Karvonen and Bas Van Heur, "Urban Laboratories: Experiments in Reworking Cities," *International Journal of Urban and Regional Research* 38, no. 2 (2014): 379–392.

81. Park, "The City as a Social Laboratory," 9.

82. Park, "The City as a Social Laboratory," 1, 15.

83. While Aristotle did not employ the term "ζώνη" (belt) in his fivefold division of the earth, the terminology of zones was subsequently attached to his schema. See Aristotle, *Meteorologica*, trans. H. D. P. Lee (Cambridge, MA: Harvard University Press, 2014), 2.5, 362a32. These terrestrial zones were defined by the angle at which the sun hit them, and each zone was understood as possessing its own particular set of conditions: the torrid zone, for example, was presumed to be uninhabitable due to its extreme heart.

84. For example, L. A. Borradaile contended in *The Animal and Its Environment* (London: Henry Frowde and Hodder and Stoughton, 1923) that *"Mountains* have not one but many faunas. A high mountain is a microcosm. As it ascends from the plain it bears zones whose climates are such as would be passed through in travelling from it to the nearer of the poles. In the tropics it will have first a tropical zone, forested or arid as the case may be; then others that are sub-tropical; temperate, with deciduous trees; cold, with grass and flowers passing into mosses and lichens; and, finally, snow-covered throughout the year" (309–310).

85. In *Chicago: An Experiment in Social Planning*, Park's colleague Ernest W. Burgess drew on the older sense of zones in his suggestion that modern cities naturally—i.e., in the absence of explicit planning—end up divided into five zones: "The Central Business District," "The Zone in Transition," "The Zone of Independent Workingmen's Homes," "The Zone of Better Residences," and "The Commuters' Zone" (114–117). The point of urban zoning plans was to replace, or at least guide, this natural zonal division of the city with zones planned by experts who drew on information gathered about the city. While early twentieth-century advocates of zoning had argued that zoning was *not* a method for enabling urban experiments—"there will be a temptation for radical individuals and officials to use zoning as a field for experimentation," Bassett noted in *Zoning*, but "this is

a mistake" (328)—it clearly was an experiment in the sense used by Park and his colleagues.

86. In the US, the 1934 Act to Provide for the Establishment, Operation, and Maintenance of Foreign-Trade Zones in Ports of Entry of the United States, to Expedite and Encourage Foreign Commerce, and for Other Purposes, H.R. 9322, Public No. 397, established the first free-trade zones. In her influential *Extrastatecraft: The Power of Infrastructure Space* (New York: Verso, 2014), Keller Easterling follows Guangwen Meng and R. J. McCalla in tracing the twentieth-century free-trade zone to a centuries-long lineage of "free" ports, which include the ancient Roman port of Delos in Greece and Hanseatic free port cities such as Hamburg. While this is in one sense correct, such a long durée account does not help us to understand why the term "zone" was specifically employed in the 1930s and misses the key fact, stressed by early twentieth-century advocates of zoning, that zones striated the unit of the cities; this was in contrast to free ports and cities, which were free in their entirety, so to speak.

87. Negroponte, *The Architecture Machine*, 185.

88. Negroponte, *The Architecture Machine*, 118.

89. In *Fundamentals of Ecology*, for example, Eugene P. Odum defined ecotones as "a junction zone or tension belt" (56; see also 57, 207). In *Imperial Eyes: Travel Writing and Transculturation* (London: Routledge, 1992), Mary Louise Pratt defined a contact zone as "the space of colonial encounters, the space in which peoples geographically and historically separated come into contact with each other and establish ongoing relations, using involving conditions of coercion, radical inequality, and intractable contact" (6); the term "is an attempt to invoke the spatial and temporal copresence of subjects previously separated but geographic and historical disjunctures, and whose trajectories now intersect" (7). In *Image and Logic* (Chicago: University of Chicago Press, 1997), Peter Galison used the term "trading zone" to get at the fact that "two groups can agree on rules of exchange even if they ascribe utterly different significance to the objects being exchanged; they may even disagree on the meaning of the exchange process itself. Nonetheless, the trading partners can hammer out a local coordination, despite vast global differences. In an even more sophisticated way, cultures in interaction frequently establish contact languages, systems of discourse that can vary from the most function-specific jargons, through semispecific pidgins, to full-fledged creoles rich enough to support activities as complex as poetry and metalinguistic reflection" (783). Latour and Peter Weibel discuss critical zones in *Critical Zones: The Science and Politics of Landing on Earth* (Cambridge, MA: MIT Press, 2020).

90. See, for example, Richard Neutra, "Comments on Planetary Reconstruction," *Arts and Architecture* 61, no. 12 (December 1944): 20–22; Neutra, "Projects of Puerto Rico: Hospitals, Health Centers, and Schools," *Architecture d'Aujourd'hui* 16, no. 5 (May 1946): 71–77; and Neutra, "Designs for Puerto Rico (A Test Case)," Richard and Dion Neutra Papers, collection no. 1,179, Department of Special Collections, Charles E. Young Research Library, University of California, Los Angeles, 1943. In these articles he labels the project a "test" and an "experiment." We thank Daniel Barber for alerting us to, and sharing, these materials.

91. Ginger Nolan, "Quasi-urban Citizenship: The Global Village as 'Nomos of the Modern,'" *Journal of Architecture* 23 (2018): 448–470. As Nolan notes, McLuhan was influenced by ethnopsychiatrist John Colin Carothers, who arrived in Kenya in the 1950s to advise British colonial authorities in their war against the Kenya Land and Freedom Army. Carothers recommended a policy of village-isation, essentially a system of forced detention in camps, as an antidote to mass collective organization in urban areas. Such ideas of a decentralized but networked system that would "buffer" denizens from the impacts of contemporary society (or its politics) inhere within contemporary fantasies of smartness as a logic to negotiate precarity and uncertainty. For critiques that stress how computing continues colonial discourses, see Kavita Philip, "The Internet Will Be Decolonized," in *Your Computer Is on Fire*, ed. Benjamin Peters, Thomas S. Mullaney, Mar Hicks, and Kavita Philip (Cambridge, MA: MIT Press, 2021), Kindle, location 2607.

92. Adam Moe Fejerskov, "The New Technopolitics of Development and the Global South as a Laboratory of Technological Experimentation," *Science, Technology, and Human Values* 42 (2017): 947–968; Kaushik Sunder Rajan, *Pharmocracy: Value, Politics, and Knowledge in Global Biomedicine* (Durham, NC: Duke University Press, 2017).

93. Ministry of Housing and Urban Affairs, Government of India, Smartnet, accessed March 9, 2022, https://smartnet.niua.org/; Nishant Shah, "Identity and Identification: The Individual in the Time of Networked Governance," *Socio-Legal Review* 11, no. 2 (2015): 22–40.

94. See Datta, "Postcolonial Urban Futures."

95. The contemporary project of so-called smart growth—also known as "new urbanism" and "sustainable development"—represents a partial approach to Negroponte's vision. Smart growth seeks to limit urban sprawl, increase use of public transportation, diversify neighborhoods, and preserve rural areas but aims to achieve these ends via economic inducements (e.g., tax credits) for development that follows these principles, rather than explicit zoning regulation. Thus, though the smartness of smart urban growth refers most explicitly to its attempts to align urban population growth with environmental and quality-of-life goals, it also references the premise that such alignment cannot be achieved by direct government regulation but only by economic inducements that are variably taken up by populations. On smart growth, see T. Chapin, "Introduction: From Growth Controls, to Comprehensive Planning, to Smart Growth: Planning's Emerging Fourth Wave," *Journal of the American Planning Association* 78, no. 1 (2012): 5–15; and J. M. DeGrove, *Planning Policy and Politics: Smart Growth and the States* (Cambridge, MA: Lincoln Institute of Land Policy, 2005).

96. Yuriko Furuhata, "Multimedia Environments and Security Operations: Expo '70 as a Laboratory of Governance," *Grey Room*, no. 54 (Winter 2014): 56–79, 59.

97. Matt Shaw, "Why Arata Isozaki Deserves the Pritzker," *The Architect's Newspaper*, April 16, 2019, https://www.archpaper.com/2019/04/arata-isozaki-pritzker/.

CHAPTER 3

1. Fischer Black, "Noise," *Journal of Finance* 41, no. 3 (1986): 529–543, 529.

2. For more on the relationship between histories of science and models of communication and economics, see Philip Mirowski, *Machine Dreams: Economics Becomes a Cyborg Science* (New York: Cambridge University Press, 2002); and Claude Shannon and Warren Weaver, *The Mathematical Theory of Communication* (1963; repr., Urbana-Champaign: University of Illinois Press, 1998). For a historical discussion about labor, work, and thermodynamics, see Anson Rabinbach, *The Human Motor: Energy, Fatigue, and the Origins of Modernity* (Berkeley: University of California Press, 1992).

3. For work on entropy and cybernetics, see N. Katherine Hayles, *How We Became Posthuman: Virtual Bodies in Cybernetics, Literature, and Informatics* (Chicago: University of Chicago Press, 1999); and Orit Halpern, "Dreams for Our Perceptual Present: Temporality, Storage, and Interactivity in Cybernetics," *Configurations* 13, no. 2 (2005): 285–321.

4. Black, "Noise," 529.

5. For an extensive discussion of thermodynamics, stochastic processes, and control, see the introduction to Norbert Wiener, *Cybernetics; or, Control and Communication in the Animal and the Machine* (Cambridge, MA: MIT Press, 1961); for further discussion, see also Halpern, "Dreams for Our Perceptual Present," and Peter Galison, "The Ontology of the Enemy: Norbert Wiener and the Cybernetic Vision," *Critical Inquiry* 21 (1994): 228–266.

6. Black was an applied mathematician who had been trained by artificial intelligence pioneer Marvin Minsky, Scholes was a Canadian American economist from the University of Chicago who came to MIT after earning his PhD under Eugene Fama, and Merton was an economist trained at MIT. The equation we discuss here is sometimes called the Black-Scholes option pricing model, while others call this the Black-Scholes derivative pricing model, and we employ the former terminology. While these three figures are not singularly responsible for global financialization, their history serves as a mirror of a situation in which new computational techniques were produced to address geopolitical-environmental transformations. See George G. Szpiro, *Pricing the Future: Finance, Physics, and the 300 Year Journey to the Black-Scholes Equation* (New York: Basic Books, 2011), 116–117, Kindle.

7. Randy Martin, "What Difference Do Derivatives Make? From the Technical to the Political Conjuncture," *Culture Unbound* 6 (2014): 189–210, 193.

8. For an account of earlier nineteenth- and twentieth-century models for pricing options, see Donald A. MacKenzie, *An Engine, Not a Camera: How Financial Models Shape Markets* (Cambridge, MA: MIT Press, 2006), 37–88.

9. Black and Scholes were not the first to treat the movement of stock prices as random, and they drew upon a tradition of research that stretched back to the early twentieth century; see MacKenzie, *An Engine, Not a Camera*, 56–66. However, they were the first to embed this premise in the mathematical structure of an investment tool that subsequently became widely used.

10. Black, "Noise," 5, cited in MacKenzie, *An Engine, Not a Camera*, 321n18.

11. Robert Merton added the concept of continuous time and figured out a derivation equation to smooth the curve of prices. The final equation is essentially the merger of a normal curve with Brownian motion; see Satyajit Das, *Traders, Guns, and Money: Knowns and Unknowns in the Dazzling World of Derivatives* (Edinburgh: Prentice Hall, 2006), 194–195.

12. MacKenzie, *An Engine, Not a Camera*, 60–67.

13. A "portfolio" is a collection of multiple investments that vary in their presumed riskiness and that aim to maximize profit for a specific level of overall risk; "arbitrage" refers to purportedly risk-free investments, such as the profit that can be made when one takes advantage of slight differences between currency exchanges—or the price of the same stock—in two different geographical locations.

14. Nick Srnicek, *Platform Capitalism* (New York: Polity Press, 2016); Shoshana Zuboff, *The Age of Surveillance Capitalism* (New York: Hachette Book Group, 2019).

15. Sarah Barns, for example, points out that the "'Uber model' isn't just for Uber, it's for any utility or infrastructure to try to adopt." Thus, "platform strategy," she argues, is an urban strategy to build digital services "in ways that ensure end users are also producers of sorts" and is part of a broader sharing economy that facilitates the intensified extraction of value from existing infrastructures. Sarah Barns, *Platform Urbanism: Negotiating Platform Ecosystems in Connected Cities* (Singapore: Palgrave Macmillan, 2020), 15. For more on Uber and platform urbanism, see Andrea Pollio, "Uber, Airports, and Labour at the Infrastructural Interfaces of Platform Urbanism," *Geoforum* 118 (January (2021): 47–55; and Arun Sundararjan, *The Sharing Economy: The End of Employment and the Rise of Crowd-Based Capitalism* (Cambridge, MA: MIT Press, 2016).

16. See Niels van Doorn, "A New Institution on the Block: On Platform Urbanism and Airbnb Citizenship," *New Media and Society* 22, no. 10 (2020): 1808–1826; Tarik Dogru, Mody Makarand, Courtney Suess, Nathan Line, and Mark Bonn, "Airbnb 2.0: Is It a Sharing Economy Platform or a Lodging Corporation?," *Tourism Management* 78 (2020): 1–5. AirBnb is also discussed at length in terms of intensified extraction by Srnicek, *Platform Capitalism*, 28, 39–44.

17. Stanislas Dehaene, *Reading in the Brain: The New Science of How We Read* (New York: Viking, 2009), 121. Dehaene explicitly cites Oliver Selfridge's Pandemonium brain model and Donald Hebb's work, both of which we discuss below.

18. Dehaene, *Reading in the Brain*, 147.

19. Johannes Bruder, *Cognitive Code: Post-Anthropocenic Intelligence and the Infrastructural Brain* (Montreal: McGill–Queen's University Press, 2020); Melissa Gregg, *Counter Productive: Time Management in the Knowledge Economy* (Durham, NC: Duke University Press, 2018).

20. On the more general role of "waste" in contemporary biomedicine, see Catherine Waldby and Robert Mitchell, *Tissue Economies: Blood, Organs, and Cell Lines in Late Capitalism* (Durham, NC: Duke University Press, 2006); on the specific biobank

examples discussed here, see Robert Mitchell and Catherine Waldby, "National Biobanks: Clinical Labour, Risk Production, and the Creation of Biovalue," *Science, Technology, and Human Values* 35, no. 3 (2010): 330–355; and Robert Mitchell, "U.S. Biobanking Strategies and Biomedical Immaterial Labor," *Biosocieties* 7, no. 3 (2012): 224–244.

21. A key element of the global order established after World War II was the linkage of oil prices to the price of the US dollar. Petrodollars had been a major source of global management, particularly in nations that were decolonizing and attempting to create autonomous economies. The surfeit of dollars played into a system that assumed that these dollars could be leveraged for energy and infrastructure development projects. Economically poorer countries were expected to take loans in US dollars in order to fund energy and resource development. The collapse of the link between dollars and oil meant that existing loans became increasingly costly (i.e., interest payments suddenly cost more in local currencies), often to the point of being impossible to pay, along with the extreme rise in costs of production and transport of goods. Despite the assumption that nations can always tax and therefore never need to default on debt, the reality proved otherwise, and massive debt default became a possibility. See Benjamin Lee and Edward LiPuma, *Financial Derivatives and the Globalization of Risk* (Durham, NC: Duke University Press, 2004).

22. Lee and LiPuma, *Financial Derivatives and the Globalization of Risk*. For more on systemic risk, see Allan R. Waldman, "OTC Derivatives and Systemic Risk: Innovative Finance or the Dance into the Abyss?," *American University Law Review* 43 (1994): 1023–1090; and Jean-Pierre Zigrand, "Systems and Systemic Risk in Finance and Economics" (SRC Special Paper, no. 1, London School of Economics and Political Science, 2014), http://eprints.lse.ac.uk/61220/1/sp-1_0.pdf.

23. Contemporary measures of GNP, GDP, and growth are all evidence of ongoing faith that growth is natural and that markets organize and provide stability to societies. Within this epistemology the idea that volatility and inefficiency are integral to market operations is anathema. For a discussion of responses to the currency situation and late 1960s volatility in markets, see Philip Mirowski, "Twelve Theses Concerning the History of Postwar Neoclassical Price Theory," *History of Political Economy* 38 (2006): 344–379, esp. 356.

24. Milton Friedman, "The Need for Futures Markets in Currencies," *Cato Journal* 31, no. 3 (2011): 635–641, 637. As MacKenzie notes, Friedman was compensated for writing this article by Leo Melamed, who was setting up an options exchange in Chicago and wanted Friedman's endorsement of such markets (*An Engine, Not a Camera*, 147).

25. Perry Mehrling, *Fischer Black and the Revolutionary Idea of Finance* (2005; repr., New York: John Wiley and Sons, 2012), 36.

26. Examples of the postwar period reworking of "reason" into algorithmic rationality include Herbert Simon, "A Behavioral Model of Rational Choice," *Quarterly Journal of Economics* 69, no. 1 (1955): 99–118; and Warren McCulloch, *Embodiments of Mind* (1965; repr., Cambridge, MA: MIT Press, 1970). For an account of these developments, see Paul Erickson et al., *How Reason Almost Lost Its Mind: The Strange Career of Cold War Rationality* (Chicago: University of Chicago Press, 2013).

27. Mehrling, *Fischer Black*, 20.

28. John von Neumann and Oskar Morgenstern, *Theory of Games and Economic Behavior*, 3rd ed. (Princeton, NJ: Princeton University Press, 1990), 8–9.

29. Lorraine Daston, "The Rule of Rules, or How Reason Became Rationality" (lecture, Max Planck Institute for the History of Science, November 21, 2010), Kindle.

30. Philip Mirowski, "Hell Is Truth Seen Too Late," *Boundary* 46, no. 1 (2019): 1–53, 3.

31. Kenneth Joseph Arrow, *Social Choice and Individual Values* (New York: Wiley, 1963), 1.

32. Arrow, *Social Choice and Individual Values*, 12.

33. There is not, to our knowledge, any critical history of optimization, and existing historical sketches written by mathematicians and economists tend to position optimization as a biological drive or natural force that received proper mathematical formulation in the eighteenth century and was more fully developed in the post–World War II period. See, for example, Ding-Zhu Du, Panos M. Pardalos, and Weili Wu, "History of Optimization," in *Encyclopedia of Optimization*, ed. Christodoulos A. Floudas and Panos M. Pardalos (New York: Springer, 2008). For a critical and nuanced account of optimization theory in economics, see Phillip Mirowski, *More Heat than Light: Economics as Social Physics; Physics as Nature's Economics* (Cambridge: Cambridge University Press, 1989); and Mirowski, *Machine Dreams*. For a discussion of optimization strategies in logistics, see Jesse LeCavalier, *The Rule of Logistics: Walmart and the Architecture of Fulfillment* (Minneapolis: University of Minnesota, 2016). As a few notes toward a future critical history of optimization, it seems that while the term "optimization" was used as early as the mid-nineteenth century, it did not—at least according to Google Ngram—enter common parlance until the 1950s. Google Ngram Viewer, s.v. "optimization," accessed November 4, 2016, https://books .google.com/ngrams/graph?content=optimization&year_start=1800&year_end=2000 &corpus=15&smoothing=3&share=&direct_url=t1%3B%2Coptimization%3B%2Cc0. A full-text search of the Institute of Electrical and Electronics Engineers Xplore Digital Library for the term "optimization" between 1890 and 1960 reveals no use of this term before 1904. There is an uptick of uses in the 1920s in articles that appear in the *Proceedings of the Institute of Radio Engineers* and then an explosion of uses in the late 1940s and 1950s, at which point the term also begins to appear in the titles of articles.

34. See, for example, William Aspray, *John von Neumann and the Origins of Modern Computing* (Cambridge, MA: MIT Press, 1990); John von Neumann, "The General and Logical Theory of Automata," in *Papers of John von Neumann on Computing and Computer Theory*, ed. William Aspray (Cambridge, MA: MIT Press, 1986); William Poundstone, *Prisoner's Dilemma* (New York: Doubleday, 1992).

35. Donald Hebb, *The Organization of Behavior: A Neuropsychological Theory* (New York: Wiley, 1949), 10–14.

36. Or, as Hebb put it, "The persistence or repetition of a reverberatory activity (or 'trace') tends to induce lasting cellular changes that add to its stability. . . . When an axon of cell *A* is near enough to excite a cell *B* and repeatedly or persistently takes part in firing it, some growth process or metabolic change takes place in one or both

cells such that A's efficiency, as one of the cells firing B, is increased" (*The Organization of Behavior*, 62).

37. According to Hebb's graduate student Woodburn Heron, the study sought to investigate the future of work conditions and, specifically, the monotony of semi- and even fully automated environments in which humans would spend hours simply overseeing machines. It was assumed that this would result in acute boredom, which would in turn lead to attention, depression, and satisfaction problems. See Woodburn Heron, "The Pathology of Boredom," *Scientific American* 196, no. 1 (January 1957): 52–57, 52–53. The study offered $20 CAD/day ($215.00 in today's US dollars) to lie in bed and do nothing. This was a huge sum, and students joked that they would lie there for a week and then could afford to go on vacation or not have to work for the rest of the year. However, most of the students lasted no more than two or three days. During the study, some individuals began having hallucinations, and cognitive functioning was often impaired. Meghan Crist, "Postcards from the Edge of Consciousness," *Nautilus*, no. 27 (August 6, 2015), http://nautil.us /issue/27/dark-matter/postcards-from-the-edge-of-consciousness-rp.

38. Simon, "A Behavioral Model of Rational Choice," 99.

39. Simon, "A Behavioral Model of Rational Choice"; see also Hunter Crowther-Heyck, *Herbert A. Simon: The Bounds of Reason in Modern America* (Baltimore: Johns Hopkins University Press, 2005); and Crowther-Heyck, *Economics, Bounded Rationality and the Cognitive Revolution* (Brookfield, VT: Edward Elgar, 1992).

40. Herbert A. Simon, *The Sciences of the Artificial* (Cambridge, MA: MIT Press, 1996), 6–7, 14, 24; for a critical contextualization of Simon's figure of the ant, see Jussi Parikka, *Insect Media* (Minneapolis: University of Minnesota Press, 2010), 135.

41. Herbert Simon, "Rational Decision Making," *American Economic Review* 69, no. 4 (1979): 493–513, 493.

42. See Lily E. Kay, "From Logical Neurons to Poetic Embodiments of Mind: Warren Mcculloch's Project in Neuroscience," *Science in Context* 14, no. 4 (2001): 591–614; and Mirowski, *Machine Dreams*, 32.

43. Our argument is not that Hebb or Simon directly influenced Black, Scholes, or Merton but rather that Hebb and Simon—along with Rosenblatt, Hayek, ecologist C. S. Holling (see chapter 4), and many other authors—helped to create a general epistemology that implicitly ratified Black's understanding of markets as noisy, which was in turn related to a more general project of developing tools for locating value in environments that could only ever be partially known. With that said, there were in fact multiple connections among Hebb, Simon, and Black: Simon, for example, was well aware of Hebb's work (see Simon's "Information-Processing Explanations of Understanding," in *The Nature of Thought: Essays in Honor of D. O. Hebb*, ed. Peter W. Jusczyk and Raymond M. Klein [Hillsdale, NJ: Lawrence Erlbaum, 1980]), and Black had intended to complete his PhD with Simon, though that did not work out; see Mehrling, *Fischer Black*, 36 (though Mehrling incorrectly identifies Simon as "Henry Simon").

44. MacKenzie, *An Engine, Not a Camera*, 123, 38.

45. Cited in Duncan Foley, "The Strange History of the Economic Agent" (paper presented to the General Seminar of the Graduate Faculty, New School University, New York, December 6, 2002), https://www.researchgate.net/publication/228563317_The _strange_history_of_the_economic_agent#:~:text=The%20curious%20history%20 of%20the,spread%20of%20capitalist%20social%20relations.&text=It%20turns%20 out%2C%20as%20well,such%20as%20physics%20and%20biology.

46. Simon, *The Sciences of the Artificial.*

47. Claudia Mareis, "Brainstorming Revisited: On Instrumental Creativity and Human Productivity in the Mid-twentieth Century," *Cultural Politics* 16, no. 1 (2020): 50–69; Horst W. J. Rittel and Melvin M. Webber, "Dilemmas in a General Theory of Planning," *Policy Sciences* 4 (1973): 155–169.

48. Orit Halpern led a research studio course organized in August 2017 through Concordia University in Montreal with Pierre-Louis Patoin from Sorbonne Nouvelle 3. This material is taken from visits to Canadian Malartic between August 2–5, 2017. Dr. Mostafa Benzaazoua, a mine reclamation expert from the Research Institute in Mining and Environment at the University of Quebec at Abitibi, was our guide and collaborator. We are grateful to his assistance and that of his colleagues in doing this research. On August 4, 2017, we were given a tour by the mine staff, whose names are being withheld as a matter of privacy.

49. The Canadian Malartic gold mine in Abitibi, Quebec, is among the largest gold mines in the world. It is certainly the largest operating in Canada and one of the largest open pits in a nation known for mining and extraction. Over half of the world's publicly listed mining and exploration companies are listed as Canadian. Mining in 2018 composed 5 percent of the nominal GDP of the nation, employed some 626,000 individuals, and was valued at CAD $260 billion. These operations are not local, as over two-thirds of the value of Canadian mining assets are located abroad, predominantly in Africa and the Americas (the United States, Mexico, and South America). Canadian Malartic gold mine staff, interviews by Orit Halpern, August 4, 2017; Canadian Malartic Partnership, accessed December 7, 2020, https:// canadianmalartic.com/en/.

50. Canadian Malartic gold mine staff, interviews.

51. These changes facilitate the expansion of the mine into deeper zones without using human labor and facilitate the construction of short interval control systems that permit all moving parts and personnel to be tracked. At this point the mine is the deepest in North America, currently dug to 3.1 kilometers underground, with hopes that automation and artificial intelligence can extend this to 3.5 kilometers underground. At such distances and depths, danger from earthquakes and earthly disruptions is sizable. Real-time analytics, improved ability to scan and map instabilities in the rock wall, and constant sensing infrastructures aim to mitigate the hazards of such terraforming projects. At the forefront of these smart mining initiatives are companies such as Cisco Systems and SAP, which are also associated with smartness in many other contemporary areas. See Joel Barde, "Canadian Mining Companies Look to 'Test Mines' to Develop New Technology," *CIM Magazine*, October 25, 2018, https://magazine.cim

.org/en/technology/a-digital-playground-en/; SAP Mining, accessed July 27, 2020, https://www.sap.com/canada/industries/mining.html.

52. Government of Canada, "Gold Facts," accessed July 29, 2020, https://www
.nrcan.gc.ca/our-natural-resources/minerals-mining/gold-facts/20514.

53. By 2002, oil markets were already the second-largest futures market and one of the largest derivatives markets across the global exchanges. See Energy Information Administration, US Department of Energy, *Derivatives and Risk Management in the Petroleum, Natural Gas, and Electricity Industries*, October 2002, http://econometricainc
.com/wp-content/uploads/2016/08/EIA_Derivatives_Report.pdf.

54. While it is unclear if this number is correct, it is what the engineers on site believed and repeated regularly.

55. The difficulty in pricing metals (and oil) as a result of derivatives markets has led Agnico Eagle to advertise that they do not sell futures. While the mine may "guarantee" the price of its gold, however, the owners of the gold clearly do not, as evidenced by the size of gold derivative and futures markets. Agnico Eagle, "60 Years in the Making," accessed December 7, 2020, https://www.agnicoeagle.com
/English/60th-anniversary/default.aspx; Canadian Malartic Partnership, "About Us," accessed December 7, 2020, http://www.canadianmalartic.com/Apropos-partenariat
-en.html.

56. Neil Brenner, "Debating Planetary Urbanization: For an Engaged Pluralism," *Environment and Planning D: Society and Space* 36, no. 3 (2018): 1–21.

57. Wikipedia, s.v. "Toronto Stock Exchange," accessed July 29, 2020, https://en
.wikipedia.org/wiki/Toronto_Stock_Exchange.

58. City of Toronto, "Innovate Here," accessed August 9, 2020, https://www.toronto
.ca/business-economy/invest-in-toronto/innovate-here/.

59. Lee and LiPuma, *Financial Derivatives and the Globalization of Risk*, 80–81.

60. See, for example, Philip Mirowski, *Machine Dreams: Economics Becomes a Cyborg Science* (New York: Cambridge University Press, 2002); and MacKenzie, *An Engine, Not a Camera*.

61. Luke Munn, "From the Black Atlantic to Black-Scholes: Precursors of Spatial Capitalization," *Cultural Politics* 16, no. 1 (2020): 92. Additional accounts of the relationships among insurance, debt, and race include Ian Baucom, *Specters of the Atlantic: A Philosophy of History* (Durham, NC: Duke University Press, 2005); Paula Chakravartty and Denise Ferreira da Silva, eds., *Race, Empire, and the Crisis of the Sub-prime* (Baltimore: Johns Hopkins University Press, 2013); Calvin Schermerhorn, *The Business of Slavery and the Rise of American Capitalism, 1815–1860* (New Haven, CT: Yale University Press, 2015); Sven Beckert, *Empire of Cotton: A Global History* (New York: Vintage Books, 2015); and Kara Keeling, *Queer Times, Black Futures* (New York: New York University Press, 2019).

62. As Lorraine Daston has noted in *Classical Probability in the Enlightenment*, eighteenth-century insurers shifted from the earlier practices of basing risk calculations on individual guesswork, sentiment, and personal relationships to using

tables, such as the one above, that enabled the automation and standardization of calculation. These latter capacities also permitted the expansion of empire and its goal of administering populations into the future.

63. While both insurance and options involve risk and both can be used to hedge bets, the two techniques are generally distinguished by the criterion that insurance only pays off in the result of the *loss* of the underlying asset (e.g., the sinking of a ship), while an option can result in profit whether the underlying asset (e.g., a company stock) goes up or down in value.

64. Munn, "From the Black Atlantic to Black-Scholes," 103.

65. Munn, "From the Black Atlantic to Black-Scholes," 105.

66. To return to our discussion in chapter 2 of urban zoning, the early twentieth-century zoning distinction between single-family housing (i.e., stand-alone houses, which were generally purchased by the inhabitant) and multifamily housing (i.e., apartment housing, which was generally rented by occupants) was itself a subtle way of enforcing racial hierarchies; see Sonia A. Hirt, "Split Apart: How Regulations Designated Populations to Different Parts of the City," in *One Hundred Years of Zoning and the Future of Cities*, ed. Amnon Lehavi (Cham, Switzerland: Springer, 2018), 3–26.

67. FICO is shorthand for a score calculated by Fair, Isaac, and Company. As Martha Poon notes, the "original Fair Isaac scorecards were custom crafted algorithmic tools designed to capture patterns of default in firm-level consumer credit data" (658n11). See Poon, "From New Deal Institutions to Capital Markets: Commercial Consumer Risk Scores and the Making of Subprime Mortgage Finance," *Accounting, Organizations and Society* 34 (2009): 654–674.

68. As Poon notes, "FICO scores can therefore be said to have reconfigured mortgage markets, putting into place a space of potential high-risk investment action. The intriguing plot twist is that these scores were introduced into the mortgage industry by risk-averse government agencies. When the GSEs adopted the FICO they interpreted scores conservatively, assuming they could be used to reinforce the binary spirit of the traditional form of credit control-by-screening. But because the tool had inscribed within it the possibility of making financially meaningful risk management calculations, it enabled the rise of a new form of financial activity: credit control-by-risk" ("From New Deal Institutions to Capital Markets," 670).

69. Aaron Glantz, *Homewreckers: How a Gang of Wall Street Kingpins, Hedge Fund Magnates, Crooked Banks, and Vulture Capitalists Suckered Millions out of Their Homes and Demolished the American Dream* (New York: HarperCollins, 2019).

70. Baucom, *Specters of the Atlantic*.

71. Keeling, *Queer Times, Black Futures*.

72. On minerals and metals markets, see Martín Arboleda, *Planetary Mine: Territories of Extraction under Late Capitalism* (New York: Verso, 2020). On energy markets, see Rusian Hharlamov and Heriner Glassbeck, "When Commodities Get Hooked on Derivatives," *Financial Times*, June 13, 2019, https://www.ft.com/content/896e47c8 -8875-11e9-a028-86cea8523dc2.

73. Greenpeace, *It's the Finance Sector, Stupid*, January 21, 2020, https://www.green peace.org/static/planet4-international-stateless/2020/01/13e3c75b-greenpeace_report _wef_2020_its-the-finance_sector_stupid.pdf.

74. Swiss Re, "Protecting Coral Reefs against Hurricane Damage," *2018 Corporate Responsibility Report*, 2018, https://reports.swissre.com/corporate-responsibility-report /2018/cr-report/solutions/strengthening-risk-resilience-2018-highlights/protecting -coral-reefs-against-hurricane-damage.html; Swiss Re, "Designing a New Type of Insurance to Protect the Coral Reefs, Economies and the Planet," news release, December 10, 2019, https://www.swissre.com/our-business/public-sector-solutions/thought -leadership/new-type-of-insurance-to-protect-coral-reefs-economies.html. See also Catrin Einhorn and Christopher Flavelle, "A Race against Time to Rescue a Reef from Climate Change," *New York Times*, December 5, 2020, https://www.nytimes.com/2020 /12/05/climate/Mexico-reef-climate-change.html.

75. Swiss Re, "Designing a New Type of Insurance."

76. For examples of the default critical account of finance, see Stephen Graham and Simon Marvin, *Splintering Urbanism: Networked Infrastructures, Technological Mobilities and the Urban Condition* (London: Routledge, 2001) (discussed in chapter 2), and most critical literature on smart cities.

77. Christopher Flavelle, "A Climate Plan in Texas Focuses on Minorities. Not Everyone Likes It," *New York Times*, July 24, 2020, https://www.nytimes.com/2020/07/24 /climate/houston-flooding-race.html. For discussion about rethinking publics, politics, infrastructures, and finance, see Stephen J. Collier, James Christopher Mizes, and Antina von Schnitzler, "Preface: Public Infrastructures/Infrastructural Publics," *Limn*, no. 7 (2016), https://limn.it/issues/public-infrastructuresinfrastructural-publics/.

78. James Christopher Mizes, "Who Owns Africa's Infrastructure?," *Limn*, no. 7 (2016), https://limn.it/issues/public-infrastructuresinfrastructural-publics/.

79. Janet Roitman, "Africa Rising: Class or Finance," podcast, University of Helsinki Seminar, 117 min., April 12, 2020, https://blogs.helsinki.fi/anthropology/2020/12 /04/janet-roitman-africa-rising-class-or-finance/.

80. Alexandra Ocasio-Cortez, H. Res. 109: Recognizing the Duty of the Federal Government to Create a Green New Deal (2019). We also take up these questions in our excursus on electrical grids in chapter 2.

81. Kenneth Boulding, "The Economics of the Coming Spaceship Earth," in *Environmental Quality in a Growing Economy*, ed. H. Jarrett (Baltimore: Resources for the Future/Johns Hopkins University Press, 1966), 3–14.

82. Martin, "What Difference Do Derivatives Make?," 189–210.

CHAPTER 4

1. Centers for Disease Control and Prevention, *Interim Pre-pandemic Planning Guidance: Community Strategy for Pandemic Influenza Mitigation in the United States—Early, Targeted, Layered Use of Nonpharmaceutical Interventions*, February 2007, https://www.cdc.gov/flu /pandemic-resources/pdf/community_mitigation-sm.pdf. According to news reports,

no one can remember who precisely wrote this report; see Mark Wilson, "The Story behind 'Flatten the Curve,' the Defining Chart of the Coronavirus," *Fast Company*, March 3, 2020, https://www.fastcompany.com/90476143/the-story-behind-flatten-the -curve-the-defining-chart-of-the-coronavirus. There are, of course, precedents for this basic approach: in the 1918 flu pandemic, infection rates in different urban areas were compared, and in the United States during World War II, similar charts demonstrated how rationing would save materials and energy for the military. We are sure there are many other examples of such precedents.

2. Lisa Parks and Janet Walker, "Disaster Media: Bending the Curve of Ecological Disruption and Moving toward Social Justice," *Media + Environment* 2, no. 1 (2020), https://mediaenviron.org/article/13474-disaster-media-bending-the-curve-of -ecological-disruption-and-moving-toward-social-justice.

3. "J. Robert Oppenheimer, Atom Bomb Pioneer, Dies," *New York Times*, February 19, 1967, https://archive.nytimes.com/www.nytimes.com/learning/general/on thisday/bday/0422.html.

4. Victor Papanek, *Design for the Real World: Human Ecology and Social Change*, 2nd ed. (Chicago: Academy Chicago Publishers, 1985), 220.

5. See the references to Neutra's work in chapter 3.

6. Quoted in Elizabeth M. DeLoughrey, "The Myth of Isolates: Ecosystem Ecologies in the Nuclear Pacific," *Cultural Geographies* 20, no. 2 (2012): 177. To this day, the Marshall Islanders continue to suffer uncompensated damage to their lives and health from the high mutation rates and ongoing cancer spawned from these tests.

7. DeLoughrey, "The Myth of Isolates." See also Richard H. Grove, *Green Imperialism: Colonial Expansion, Tropical Island Edens and the Origins of Environmentalism, 1600–1860* (Cambridge: Cambridge University Press, 1996).

8. Joel B. Hagen, *An Entangled Bank: The Origins of Ecosystem Ecology* (New Brunswick, NJ: Rutgers University Press, 1992), 55–62, 64–68.

9. Hagen, *An Entangled Bank*, 65.

10. Peder Anker, *Imperial Ecology: Environmental Order in the British Empire, 1895–1945* (Cambridge, MA: Harvard University Press, 2001), 31.

11. Anker, *Imperial Ecology*.

12. Howard T. Odum and Eugene P. Odum, "Trophic Structure and Productivity of a Windward Coral Reef Community on Enitwetok Atoll," *Ecological Monographs* 25 (July 1955): 291.

13. Laura J. Martin, "The X-ray Images That Showed Midcentury Scientists How Radiation Affects an Ecosystem," *Slate*, December 28, 2015, https://slate.com/human -interest/2015/12/how-midcentury-ecologists-used-x-ray-radioautographs-to-see -how-radiation-moves-through-bodies.html.

14. Odum apparently attended the conferences once as well but thought they were a waste of time (Hagen, *An Entangled Bank*, 213). However, he is also considered by many historians and documentarists, including Adam Curtis (*All Watched Over by Machines of Loving Grace*, BBC Films, 2011) and Peter J. Taylor, as being central to

introducing cybernetic concepts and postwar approaches to systems, feedback and self-regulation, and organization into ecology. See Peter J. Taylor, *Unruly Complexity: Ecology, Interpretation, Engagement* (Chicago: University of Chicago Press, 2010).

15. G. E. Hutchinson, "Circular Causal Systems in Ecology," *Annals of the New York Academy of Science* 40 (1948): 221–246; for context, see Brian Lindseth, "The Pre-history of Resilience in Ecological Research," *Limn*, no. 1 (2011), https://limn.it /articles/the-pre-history-of-resilience-in-ecological-research/.

16. Peter J. Taylor, "Technocratic Optimism, H.T. Odum, and the Partial Transformation of Ecological Metaphor after World War II," *Journal of the History of Biology* 21 (1988): 217.

17. Howard T. Odum, *Systems Ecology: An Introduction*, ed. Robert L. Metcalf and Werner Stumm (New York: John Wiley and Sons, 1983), ix–x.

18. His brother Eugene summed up this approach in his foundational text to the field, *Ecology* (New York: Holt, Rinehart, and Winston, 1963, repr. 1975): "A very important corollary . . . is the principle of hierarchical control. Simply stated, this principle is as follows: As components combine to produce larger functional wholes in a hierarchical series, new properties emerge . . . to understand and properly manage a forest we must not only be knowledgeable about trees as populations, but we must also study the forest as an ecosystem" (5).

19. On the importance of Lotka for both Hutchison and Odum, see Taylor, "Technocratic Optimism," 217, 219, 225–226. On the differences between Lotka's and Mayr's approaches to animal populations, see Philip Kreager, "Darwin and Lotka: Two Concepts of Population," *Demographic Research* 21, no. 2 (2009): 469–502.

20. Taylor, "Technocratic Optimism."

21. For more on this, see Orit Halpern, *Beautiful Data: A History of Vision and Reason since 1945* (Durham, NC: Duke University Press, 2015); and Halpern, "Dreams for Our Perceptual Present: Temporality, Storage, and Interactivity in Cybernetics," *Configurations* 13, no. 2 (2005): 283–319; see also Paul N. Edwards, *The Closed World: Computers and the Politics of Discourse in Cold War America* (Cambridge, MA: MIT Press, 1996).

22. See E. Odum, *Ecology*, 7–8, 152.

23. "Agent Orange," History Channel, May 16, 2019, https://www.history.com /topics/vietnam-war/agent-orange-1. See also Jacob Darwin Hamblin, *Arming Mother Nature: The Birth of Catastrophic Environmentalism* (Oxford: Oxford University Press, 2013).

24. Donella H. Meadows et al., *The Limits to Growth: A Report for the Club of Rome's Project on the Predicament of Mankind* (New York: Universe Books, 1972), esp. 1–54.

25. The Volkswagen Foundation gave USD $250,000, and a group was assembled at MIT to prepare a comprehensive model. Five sectors were chosen—demography, economics, geology, ecology, and agriculture—and each area was studied and the findings added to the basic model structure from Forrester. In 1971 the Club of Rome conceived that this research should become a report, *The Limits to Growth*.

See Fernando Elichirigoity, *Planet Management: Limits to Growth, Computer Simulation, and the Emergence of Global Spaces* (Evanston, IL: Northwestern University Press, 1999).

26. As Elichirigoity notes in *Planet Management*, Forrester himself did not end up personally leading the *Limits to Growth* team, as some of the study's funders had expected, because Forrester had already committed himself to another project of computationally modeling a national economy (95).

27. Donella H. Meadows, "The History and Conclusions of *Limits to Growth*," *System Dynamics Review* 23 (2007): 191–197, 193.

28. Elichirigoity, *Planet Management*, 66, 78–77.

29. For a helpful history of the "closing" of the world in this way, see Edwards, *The Closed World*.

30. See Arturo Escobar, *Encountering Development: The Making and Unmaking of the Third World*, 2nd ed. (Princeton, NJ: Princeton University Press, 2012); Michelle Murphy, *Seizing the Means of Reproduction* (Durham, NC: Duke University Press, 2012); Thomas Robertson, "'This Is the American Earth': American Empire, the Cold War, and American Environmentalism," *Diplomatic History* 32, no. 4 (2008): 561–584.

31. On the emergence in the 1950s of economic "growth" as an agreed-upon international goal of Western democratic countries, see Scott O'Bryan, *The Growth Idea: Purpose and Prosperity in Postwar Japan* (Honolulu: University of Hawai'i Press, 2009), 133–171; Robert M. Collins, *More: The Politics of Economic Growth in Postwar America* (Oxford: Oxford University Press, 2000), 17–39; David W. Ellwood, *Rebuilding Europe: Western Europe, America, and Postwar Reconstruction* (London: Longman, 1992), 205–241; and Matthias Schmelzer, *The Hegemony of Growth: The OECD and the Making of the Economic Growth Paradigm* (Cambridge: Cambridge University Press, 2016), esp. 142–162. As Schmelzer notes, one important impetus for this emphasis on growth was the perception that the economies of the USSR and its allies were expanding at a rate of 10–15 percent per year (151).

32. Friedrich Hayek, "The Pretence of Knowledge" (lecture, Sveriges Riksbank Prize in Economic Sciences in Memory of Alfred Nobel, December 11, 1974), https://www.nobelprize.org/prizes/economic-sciences/1974/hayek/lecture/.

33. Hayek, "The Pretence of Knowledge."

34. Milton Friedman, "The Need for Futures Markets in Currencies," *Cato Journal* 31, no. 3 (2011): 635–641, 636. This article is a reprint of a December 20, 1971, report to the Chicago Mercantile Exchange, and as Donald MacKenzie documents, this was an "article for hire," as the head of the Chicago Mercantile Exchange paid Friedman to write the article, which was intended to (and did) pave the way for federal approval of precisely such a market. Hence, the article advocates not only for a futures currency market but for its location in the United States. Friedman contended that "as Britain has demonstrated in the nineteenth century, financial services of all kinds can be a highly profitable export commodity," and he proposed that a US-based futures market would strengthen the American position while also maintaining the stability and expansion of global trade.

35. Friedman, "The Need for Futures Markets," 637. Friedman stressed that this market "cannot depend solely on hedging transactions by persons involved in foreign trade and investment." In addition, he said that the "market needs speculators who are willing to take open positions as well as hedges. The larger the volume of speculative activity, the better the market and the easier it will be for persons involved in foreign trade and investment to hedge at low costs and at market prices that move only gradually and are not significantly affected by even a large commercial transactions" (638). The terminology of resilience seems not to have been Friedman's innovation, as other economists had also used this term in the late 1960s when discussing the need for Bretton Woods reform.

36. Jeremy Walker and Melinda Cooper also connect Holling and Hayek in "Genealogies of Resilience: From Systems Ecology to the Political Economy of Crisis Adaptation," *Security Dialogue* 14, no. 2 (2011): 143–160, and we draw on some of their analysis here.

37. C. S. Holling, "Resilience and Stability of Ecological Systems," *Annual Review of Ecological Systems* 4 (1973): 1–23, 1.

38. Peter Larkin, *The Canadian Encyclopedia*, s.v. "Crawford Stanley Holling," last modified July 11, 2020, https://thecanadianencyclopedia.ca/en/article/crawford-stanley-holling.

39. C. S. Holling, "The Spruce Budworm/Forest Management Problem," in *Adaptive Environmental Assessment and Management*, ed. C. S. Holling (New York: John Wiley and Sons, 1978), 143–183.

40. For a summary of strategies in adaptive management, see Holling, *Adaptive Environmental Assessment and Management*. The emphasis of Holling and his coauthors in *Adaptive Environmental Assessment and Management* on small teams and regional (vs. global) projects was presumably also intended as a critique of the "big science" approach that characterized the International Biological Program (IBP; 1964–1974), one part of which was headed by Howard Odum. On big science and the IBP, see Hagen, *An Entangled Bank*, 165–189.

41. Holling, "Resilience and Stability of Ecological Systems," 21. Holling in fact distinguished between two kinds of resilience: "engineering resilience" and "ecological resilience." Engineering resilience is mechanical and incapable of rapid evolutionary change, and, within ecology, its practices are indebted to ideas of perfectly recording the world, counting populations, and assuming that the future can be predicted directly from past data. Ecological resilience contests each of these premises. See especially Holling's essay "Engineering Resilience versus Ecological Resilience," in *Engineering within Ecological Constraints*, ed. Peter C. Schulze (Washington, DC: National Academies Press, 1996), 31–44, reprinted in Lance H. Gunderson, Craig R. Allen, and C. S. Holling, eds., *Foundations of Ecological Resilience* (Washington, DC: Island Press, 2009), 51–66.

42. For a summary of strategies in adaptive management, see Holling, *Adaptive Environmental Assessment and Management*.

43. Holling, "Resilience and Stability of Ecological Systems," 21.

44. Holling, *Adaptive Environmental Assessment and Management*, 9.

45. This observation about Gaia is inspired by Melinda Cooper, *Life as Surplus* (Seattle: University of Washington Press, 2008), 35.

46. Burton G. Malkiel and Eugene F. Fama, "Efficient Capital Markets: A Review of Theory and Empirical Work," *Journal of Finance* 25, no. 2 (1970): 383–417.

47. We might also add those social theorists and historians who introduced models of abrupt shifts of either paradigms (Thomas Kuhn) or epistemes (Michel Foucault).

48. Or, as Holling and his coauthors put it, "Human systems have the same four properties (namely, organized connection between parts, spatial heterogeneity, resilience, and dynamic variability) that ecological systems have" (35–36).

49. See the work by Holling and others on a panarchy model of ecology in Lance H. Gunderson and C. S. Holling, eds., *Panarchy: Understanding Transformations in Human and Natural Systems* (Washington, DC: Island Press, 2002).

50. For the many discourses into which Holling-inspired concepts of resilience have spread, see Melinda Cooper and Jeremy Walker, "Genealogies of Resilience: From Systems Ecology to the Political Economy of Crisis Adaptation," *Security Dialogue* 14, no. 2 (2011): 143–160; Andreas Folkers, "Continuity and Catastrophe: Business Continuity Management and the Security of Financial Operations," *Economy and Society* 46, no. 1 (2017): 103–127; Kevin Grove, *Resilience* (New York: Routledge, 2018); David Chandler, "Resilience and Human Security: The Post-Interventionist Paradigm," *Security Dialogue* 43, no. 3 (2012): 213–229; Chandler, *Resilience: The Governance of Complexity* (New York: Routledge, 2014); and Stockholm Resilience Centre, "About Us," Stockholm University, 2020, https://www.stockholmresilience.org/about-us.html. For work by Holling and the Resilience Alliance, see C. S. Holling, *Adaptive Environmental Assessment and Management* (New York: John Wiley and Sons, 1978); "Ontario Completes Sale of 14 Million Shares of Hydro One to Ontario First Nations," *NetNewsLedger*, January 2, 2018.

51. Stockholm Resilience Centre, https://www.stockholmresilience.org/about-us.html.

52. Holling, *Adaptive Environmental Assessment and Management*, 3–4.

53. Holling, *Adaptive Environmental Assessment and Management*, ix–x.

54. Holling, *Adaptive Environmental Assessment and Management*, xi.

55. See, e.g., Holling, *Adaptive Environmental Assessment and Management*, 215–242.

56. For a history of scenario planning within Cold War policy circles and corporations (especially RAND), see Bradfield et al., "The Origins and Evolution of Scenario Techniques in Long Range Business Planning," *Futures* 37 (2005): 795–812. As Bradfield et al. note, Herman Kahn's coauthored *The Year 2000: A Framework for Speculation on the Next Thirty-Three Years* (1967) helped to popularize the concept and practice of planning by means of future "scenarios" (799).

57. Meadows et al., *The Limits to Growth*, 93, 186.

58. Meadows et al., *The Limits to Growth*, 121.

59. See Folkers, "Continuity and Catastrophe," 103–127; Folkers, *Das Sicherheitsdispositiv der Resilienz: Katastrophische Risiken und die Biopolitik vitaler Systeme* (Frankfurt am Main: Campus, 2018).

60. Folkers, "Continuity and Catastrophe," 104–105.

61. Holling uses the phrase "learning with continuity" in "Understanding the Complexity of Economic, Ecological, and Social Systems," *Ecosystems* 4, no. 5 (2001): 390–405, at 390, 399, 402.

62. Goldman Sachs Global Business Continuity Program, *Business Continuity and Technology Resilience Program for Disaster Recovery: Overview*, 2020, accessed December 29, 2020, https://www.goldmansachs.com/disclosures/business-continuity.pdf.

63. See Folkers, "Continuity and Catastrophe"; Stephen Collier and Andrew Lakoff, "Vital Systems Security: Reflexive Biopolitics and the Government of Emergency," *Theory, Culture and Society* 32, no. 3 (2015): 19–51; and Deborah Cowen, *The Deadly Life of Logistics* (Minneapolis: University of Minnesota Press, 2014). For more on the Cold War legacy of securitization and vital systems security, see Joseph Masco, *The Theater of Operations: National Security Affect from the Cold War to the War on Terror* (Durham, NC: Duke University Press, 2014); Stephen Collier and Andrew Lakoff, "Introduction: Systemic Risk," *Limn*, no. 1 (2011). For documents on vital system security policy in the US government and finance, see Barak Obama, the White House, Office of the Press Secretary, *National Security Strategy*, February 6, 2015, https://obamawhitehouse.archives.gov/sites/default/files/docs/2015_national_secu rity_strategy_2.pdf; Goldman Sachs Global Business Continuity Program, *Business Continuity and Technology Resilience Program*.

64. Nassim Taleb, *Antifragile: Things That Gain from Disorder* (New York: Random House, 2012), 67.

65. Taleb, *Antifragile*, 4–5.

66. Debjani Bhattacharyya, *Empire and Ecology in the Bengal Delta: The Making of Calcutta* (Cambridge: Cambridge University Press, 2018), Kindle.

67. Suman Chakraborti, "Data Maturity: New Town Is India's 8th Smartest City," *Times of India*, June 29, 2021, https://timesofindia.indiatimes.com/city/kolkata/data -maturity-new-town-is-indias-8th-smartest-kolkata/articleshow/83943513.cms.

68. Sedeshna Mitra, "Roads to New Urban Futures: Flexible Territorialisation in Peri-urban Kolkata and Hyderbad," *Economic and Political Weekly* 53, no. 49 (2018): 56–64; Srestha Banerjee, "An Evaluation of the Political Economy of Urban Ecological Sustainability in Indian Cities in a Globalizing Era: A Perspective from the East Kolkata Wetlands" (PhD diss., University of Delaware, 2009).

69. While actual information about the financing of the Rajarat development is difficult to find, data about the lack of occupancy and the probable financing came from discussions with the Kolkata Research Group members Ranabir Samadar and Mithelesh Kumar on March 31 and April 1, 2016. For further research on the zone produced by the Transit Labour project, directed by Brett Neilson and Ned Rossiter, see Transit Labour, "Platform: Kolkata," accessed April 8, 2017, http://transitlabour .asia/kolkata/.

70. Sudeshna Banerjee, "The March of the Mega-city: Governance in West Bengal and the Wetlands to the East of Kolkata," *South Asia Chronicle* 2 (2012): 93–118; Patrick Barkham, "The Miracle of Kolkata's Wetlands—and One Man's Struggle to Save Them," *Guardian*, March 9, 2016; Sana Huque, Sarmistha Pattanaik, and D. Parthasarathy, "Cityscape Transformation and the Temporal Metamorphosis of East Kolkata Wetlands: A Political Ecology Perspective," *Sociological Bulletin* 69, no. 1 (2020): 102–103.

71. Dhrubajyoti Ghosh, *Ecology and Traditional Wetland Practice: Lessons from Wastewater Utilisation in the East Calcutta Wetlands* (Kolkata: Worldview, 2005).

72. Ministry of Urban Development, Government of India, *Smart Cities: Mission Statement and Guidelines*, June 2015, http://smartcityrajkot.in/Docs/SmartCityGuidelines .pdf.

73. Ministry of Urban Development, *Smart Cities*.

74. Paul Virilio, *Speed and Politics* (Cambridge, MA: MIT Press, 2006).

75. See, for example, the *New York Times* series on resilience in the COVID-19 era, May 3, 2021, https://www.nytimes.com/spotlight/resilience.

76. This is also another example of that weaponization of scientific uncertainty called "agnatology" by historians of science Robert Proctor and Naomi Oreske and that is characteristic of the business strategies of what Oreske calls "merchants of doubt." Proctor and Oreske documented the deliberate use, and cultivation, of scientific uncertainty by tobacco and energy companies as a means of preventing or delaying government regulation, and Philip Mirowski has noted that this strategy is part of a more general neoliberal attack on independent (i.e., noncorporate) scientific knowledge. Within the terms of our chapter, agnatology names methods of profiting from the uncertainty inherent in complex systems, and this project is of course facilitated by the premise that error is intrinsic to every attempt to model that complexity. In the case of COVID-19, the more general strategy of agnatology also resonated with President Trump's authoritarian strategies. Trump's political career was founded on a critique of elitism and scientific forms of evidence and evidence-based decision-making. The early uncertainty within scientific forums about COVID-19 facilitated this critique of experts and science and allowed the Right to transform the catastrophe into a war of ideologies. Within this war of ideologies, Trump could effectively position his authoritarian confidence as the best and most valid voice while simultaneously suggesting that some people should be sacrificed for the economy. See Nicholas B. King, "Briefing: Evidence and Uncertainty during the Covid-19 Pandemic," Max Bell School of Public Policy, McGill University, https://www.mcgill.ca/maxbellschool/article/briefing-evidence-and-uncertainty-during -covid-19-pandemic. David Chandler has also argued that resilience and concepts of complexity have become excuses for government not to interfere in situations of crisis or disaster. Chandler, *Resilience: The Governance of Complexity*.

77. Kara Keeling, *Queer Times, Black Futures* (New York: New York University Press, 2019), 22.

78. Keeling, *Queer Times, Black Futures*, 163.

79. Ferris Jabr, "The Social Life of Forests," *New York Times*, December 12, 2020.

80. Suzanne W. Simard et al., "Net Transfer of Carbon between Ectomycorrhizal Tree Species in the Field," *Nature* 388 (August 7, 1997): 579–582; Simard, "Interspecific Carbon Transfer in Ectomycorrhizal Tree Species Mixtures" (PhD diss., Oregon State University, 1995).

81. Suzanne W. Simard, "Nature's Internet: How Trees Talk to Each Other in a Healthy Forest," 2017, TEDxSeattle, February 2, 2017, https://www.youtube.com /watch?v=breDQqrkikM.

82. Jabr, "The Social Life of Forests"; Brandon Keim, "Never Underestimate the Intelligence of Trees," *Nautilus*, October 30, 2019, https://nautil.us/never-underestimate -the-intelligence-of-trees-8573/.

83. Elise Filotas et al., "Viewing Forests through the Lens of Complex Systems Science," *Ecosphere* 5, no. 1 (2014): 1–23.

84. Filotas et al., "Viewing Forests."

85. Keeling, *Queer Times, Black Futures*, 5.

86. Keeling, *Queer Times, Black Futures*, xiv–xv, 153–155. On the Black radical tradition, see Cedric J. Robinson, *Black Marxism: The Making of the Black Radical Tradition* (1986; repr., Chapel Hill: University of North Carolina Press, 2000); on freedom dreams, see R. D. G. Kelley, *Freedom Dreams: The Black Radical Imagination* (Boston: Beacon Press, 2002).

87. Stefano Harney and Fred Moten, *The Undercommons: Fugitive Planning and Black Study* (New York: Minor Compositions, 2013), 151.

CODA

1. See, for example, Edward Nik-Khah, "George Stigler, the Graduate School of Business, and the Pillars of the Chicago School," in *Building Chicago Economics: New Perspectives on the History of America's Most Powerful Economics Program*, ed. Robert Van Horn, Philip Mirowski, and Thomas A. Stapleford (Cambridge: Cambridge University Press, 2011), 116–147; Edward Nik-Khah, "The 'Marketplace of Ideas' and the Centrality of Science to Neoliberalism," in *The Routledge Handbook of the Political Economy of Science*, ed. David Tyfield, Rebecca Lave, Samuel Randalls, and Charles Thorpe (London: Routledge, 2017), 31–42; Philip Mirowski, "Hell Is Truth Seen Too Late," *boundary 2* 46, no. 1 (2019): 1–53. See also Wendy Brown, *Undoing the Demos: Neoliberalism's Stealth Revolution* (New York: Zone Books, 2015).

2. See Nik-Khah, "George Stigler" and "The 'Marketplace of Ideas'"; Brown, *Undoing the Demos*.

3. See Carl Schmitt, *The Crisis of Parliamentary Democracy*, trans. Ellen Kennedy (Cambridge, MA: MIT Press, 1985); Leo Strauss, *Natural Right and History* (Chicago: University of Chicago Press, 1953), 6.

4. See Jonathan Mattingly, "Declaration of Jonathan Mattingly," Common Cause, et al. v. League of Women Voters of North Carolina, et al. Civil Action No.

1:16-CV-1026-WO-JEP/Civil Action No. 1:16-Cv-1164-WO=JEP (US District Court for the Middle District of North Carolina, March 6, 2017), https://s10294.pcdn.co/wp-content/uploads/2016/05/Expert-Report-of-Jonathan-Mattingly.pdf.

5. Joseph Dumit, "Circuits in the Mind" (lecture, New York University, 2007).

6. The so-called Caravan of Death carried out the executions of 26 people in Chile's south desert and 71 in the north desert. BBC News, "Chile's Caravan of Death: Ex-Army Chief Cheyre Convicted for Pinochet-Era Crimes," November 10, 2018, https://www.bbc.com/news/world-latin-america-46160437. See also National Commission for Truth and Reconciliation, "Report of the Chilean National Commission on Truth and Reconciliation" (Santiago, 1990), https://web.archive.org/web/20091124213320/http://www.usip.org/resources/truth-commission-chile-90. Orit Halpern wishes to thank Katheryn Detwiler for her research assistance and aid in doing this Chile research. Her dissertation, "Science Fell in Love with the Chilean Sky: Data as a Speculative Resource in the Atacama Desert," explores connections between astronomy and extraction in the Atacama Desert. I am indebted to her labor in helping me gain access to the ALMA Observatory, the Soquimich lithium mine, and the CMM laboratory, and I appreciate the productive discussions ensuing from the research concerning data, time, landscapes, and trauma in this desert.

BIBLIOGRAPHY

Abramson, Daniel M. *Obsolescence: An Architectural History*. Chicago: University of Chicago Press, 2016.

Agnico Eagle. "60 Years in the Making." Accessed December 7, 2020. https://www.agnicoeagle.com/English/60th-anniversary/default.aspx.

Alexander, D. E. "Resilience and Disaster Risk Reduction: An Etymological Journey." *Natural Hazards and Earth System Sciences* 13 (2013): 2707–2716.

Allen, Krista. "NTUA, Kayenta Solar Project Chart Path to the Future." *Navajo Times*, October 3, 2019. https://navajotimes.com/biz/ntua-kayenta-solar-project-chart-path-to-the-future/.

Amin, Ash. "On Urban Failure." *Social Research* 83, no. 3 (2016): 778.

Amoore, Louise. "Data Derivatives: On the Emergence of a Security Risk Calculus for Our Times." *Theory, Culture and Society* 28, no. 6 (2011): 24–43.

Anderson, Jannine. "Navajo Solar Project Sending Power to the Grid in Kayenta, Ariz." *American Public Power Association*, June 20, 2017. https://www.publicpower.org/periodical/article/navajo-solar-project-sending-power-grid-kayenta-ariz.

Anker, Peder. *From Bauhaus to Ecohouse: A History of Ecological Design*. Baton Rouge: Louisiana State University Press, 2010.

Anker, Peder. *Imperial Ecology: Environmental Order in the British Empire, 1895–1945*. Cambridge, MA: Harvard University Press, 2001.

Arboleda, Martín. *Planetary Mine: Territories of Extraction under Late Capitalism*. New York: Verso, 2020.

Aristotle. *Meteorologica*. Translated by H. D. P. Lee. Cambridge, MA: Harvard University Press, 2014.

Arrow, Kenneth Joseph. *Social Choice and Individual Values*. New York: Wiley, 1963.

Aseltine, J. A., A. Mancini, and C. Sarture. "A Survey of Adaptive Control Systems." *IRE Transactions on Automatic Control* 6, no. 1 (1958): 102–108.

Aspray, William. *John von Neumann and the Origins of Modern Computing*. Cambridge, MA: MIT Press, 1990.

Bacon, Francis. *The Essays, or Councils, Civil and Moral, of Sir Francis Bacon, Lord Verulam, Viscount St. Alban with a Table of the Colours of Good and Evil, and a Discourse of the Wisdom of the Ancients: To This Edition Is Added the Character of Queen Elizabeth, Never before Printed in English*. London: Printed for H. Herringman, R. Scot, R. Chiswell, A. Swalle, and R. Bentley, 1696.

Banerjee, Srestha. "An Evaluation of the Political Economy of Urban Ecological Sustainability in Indian Cities in a Globalizing Era: A Perspective from the East Kolkata Wetlands." PhD diss., University of Delaware, 2009.

Banerjee, Sudeshna. "The March of the Mega-city: Governance in West Bengal and the Wetlands to the East of Kolkata." *South Asia Chronicle* 2 (2012): 93–118.

Barber, Daniel. *Modern Architecture and Climate: Design before Air Conditioning*. Princeton, NJ: Princeton University Press, 2020.

Barde, Joel. "Canadian Mining Companies Look to 'Test Mines' to Develop New Technology." *CIM Magazine*, October 25, 2018. https://magazine.cim.org/en/technology/a-digital-playground-en/.

Barkham, Patrick. "The Miracle of Kolkata's Wetlands—and One Man's Struggle to Save Them." *Guardian*, March 9, 2016. https://www.theguardian.com/cities/2016/mar/09/kolkata-wetlands-india-miracle-environmentalist-flood-defence#:~:text=The%20wetlands%20produce%2010%2C000%20tonnes,so%20close%20to%20the%20centre.

Barns, Sarah. *Platform Urbanism: Negotiating Platform Ecosystems in Connected Cities*. Singapore: Palgrave Macmillan, 2020.

Bassett, Edward Murray. *Zoning*. 1922. Reprint, New York: Russell Sage Foundation, 1940.

Baucom, Ian. *Specters of the Atlantic: A Philosophy of History*. Durham, NC: Duke University Press, 2005.

BBC News. "Chile's Caravan of Death: Ex-Army Chief Cheyre Convicted for Pinochet-Era Crimes." November 10, 2018. https://www.bbc.com/news/world-latin-america-46160437.

Beckert, Sven. *Empire of Cotton: A Global History*. New York: Vintage Books, 2015.

Beiler, Kevin J., Daniel M. Durall, Suzanne W. Simard, Sheri A. Maxwell, and Annette M. Kretzer. "Architecture of the Wood-Wide Web: *Rhizopogon* spp. Genets Link Multiple Douglas-Fir Cohorts." *New Phytologist* 185, no. 2 (2010): 543–553.

Bell, Terence. "An Overview of Commercial Lithium Production." Thought. Co. Last modified August 21, 2020. https://www.thoughtco.com/lithium-production-2340123.

Bellamy, Paul B. *A History of Workmen's Compensation, 1898–1915: From Courtroom to Boardroom.* New York: Garland Publishing, 1997.

Bendix, Aria. "New York's $25 Billion Megadevelopment Can Withstand a Superstorm or Terrorist Attack—Even If the Entire City Shuts Down." Business Insider, March 14, 2019. https://www.businessinsider.com/hudson-yards-can-withstand-superstorm -terrorist-attack-2019-3.

Benson, Etienne. *Surroundings: A History of Environments and Environmentalisms.* Chicago: University of Chicago Press, 2020.

Bergdoll, Barry. "Introductory Statement." Museum of Modern Art. Accessed November 3, 2016. https://www.moma.org/explore/inside_out/rising-currents?x-iframe=true #description. The original website has disappeared, but similar materials can be accessed at https://www.moma.org/calendar/exhibitions/1028.

Berry, Michael W., and Murray Browne. *Understanding Search Engines: Mathematical Modeling and Text Retrieval.* Philadelphia: Society for Industrial and Applied Mathematics, 2005.

Bhattacharyya, Debjani. *Empire and Ecology in the Bengal Delta: The Making of Calcutta.* Cambridge: Cambridge University Press, 2018. Kindle.

Black, Fischer. "Noise." *Journal of Finance* 41, no. 3 (1986): 529–543.

Blanchard, Elodie Vieille. "Modelling the Future: An Overview of the 'Limits to Growth' Debate." *Centaurus* 52 (2010): 91–116.

Borradaile, L. A. *The Animal and Its Environment.* London: Henry Frowde and Hodder and Stoughton, 1923.

Boulding, Kenneth. "The Economics of the Coming Spaceship Earth." In *Environmental Quality in a Growing Economy*, edited by H. Jarrett, 3–14. Baltimore: Resources for the Future/Johns Hopkins University Press, 1966.

Box, G. E. P. "Evolutionary Operation: A Method for Increasing Industrial Productivity." *Applied Statistics* 6, no. 2 (1957): 81–101.

Boyle, Rebecca. "The Search for Alien Life Begins in Earth's Oldest Desert." *Atlantic*, November 28, 2018. https://www.theatlantic.com/science/archive/2018/11/searching -life-martian-landscape/576628/.

Bradfield, R., George Wright, George Burt, George Cairns, and Kees Van Der Heijden. "The Origins and Evolution of Scenario Techniques in Long Range Business Planning." *Futures* 37 (2005): 795–812.

Brandt, Marisa Renee. "War, Trauma, and Technologies of the Self: The Making of Virtual Reality Exposure Therapy." PhD diss., University of San Diego, 2013.

Bratton, Benjamin. *The Stack: On Software and Sovereignty*. Cambridge, MA: MIT Press, 2016.

Brenner, Neil. "Debating Planetary Urbanization: For an Engaged Pluralism." *Environment and Planning D: Society and Space* 36, no. 3 (2018): 1–21.

Brilliant.org. "Black-Scholes-Merton." Accessed June 20, 2020. https://brilliant.org/wiki /black-scholes-merton/.

Brin, Sergey, and Lawrence Page. "The Anatomy of a Large-Scale Hypertextual Web Search Engine." Working paper, Stanford InfoLab, Stanford University. Accessed March 6, 2016. http://infolab.stanford.edu/~backrub/google.html.

Brown, Wendy. *Undoing the Demos: Neoliberalism's Stealth Revolution*. New York: Zone Books, 2015.

Bruder, Johannes. *Cognitive Code: Post-Anthropocenic Intelligence and the Infrastructural Brain*. Montreal: McGill–Queen's University Press, 2020.

Burgess, Ernest W. "Urban Areas." In *Chicago: An Experiment in Social Science Research*, edited by Thomas Vernor Smith and Leonard Dupee White, 113–138. Chicago: University of Chicago Press, 1929.

Busbea, Larry D. *The Responsive Environment: Design Aesthetics and the Human in the 1970's*. Minneapolis: University of Minnesota Press, 2019.

Caldwell, Bruce. Introduction to F. A. Hayek, *The Road to Serfdom: Text and Documents*. Edited by Bruce Caldwell. Chicago: University of Chicago Press, 2007, 1–33.

Canadian Malartic Partnership. "About Us." Accessed December 7, 2020. http://www .canadianmalartic.com/Apropos-partenariat-en.html.

Carr, Nicholas. *The Big Switch: Rewiring the World, from Edison to Google*. New York: W. W. Norton, 2008.

Carrere, Michelle. "Chile Renews Contract with Lithium Company Criticized for Damaging Wetland." Translated by Sydney Sims. *Mongabay*, December 16, 2018. https://news.mongabay.com/2018/12/chile-renews-contract-with-lithium-company -criticized-for-damaging-wetland/.

Castells, Manuel. *The Rise of the Network Society*. New York: Wiley-Blackwell, 2000.

Centers for Disease Control and Prevention. *Interim Pre-pandemic Planning Guidance: Community Strategy for Pandemic Influenza Mitigation in the United States—Early, Targeted, Layered Use of Nonpharmaceutical Interventions*. Atlanta: Centers for Disease Control and Prevention, 2007.

Chakraborti, Suman. "Data Maturity: New Town Is India's 8th Smartest City." *Times of India*, June 29, 2021. https://timesofindia.indiatimes.com/city/kolkata/data -maturity-new-town-is-indias-8th-smartest-kolkata/articleshow/83943513.cms.

Chakravartty, Paula, and Denise Ferreira da Silva, eds. *Race, Empire, and the Crisis of the Subprime Baltimore*. Baltimore: Johns Hopkins University Press, 2013.

Chandler, David. *Resilience: The Governance of Complexity*. New York: Routledge, 2014.

Chandler, David. "Resilience and Human Security: The Post-Interventionist Paradigm." *Security Dialogue* 43, no. 3 (2012): 213–229.

Chang, Lian Chikako. "Live Blog—Mohsen Mostafavi in Conversation with Nicholas Negroponte." *Archinect* (blog), January 15, 2013. https://archinect.com/lian/live-blog-mohsen-mostafavi-in-conversation-with-nicholas-negroponte.

Chapin, T. "Introduction: From Growth Controls, to Comprehensive Planning, to Smart Growth: Planning's Emerging Fourth Wave." *Journal of the American Planning Association* 78, no. 1 (2012): 5–15.

City of Toronto. "Innovate Here." Accessed August 9, 2020. https://www.toronto.ca/business-economy/invest-in-toronto/innovate-here/.

Collier, Stephen, and Andrew Lakoff. "Introduction: Systemic Risk." *Limn*, no. 1 (2011). https://limn.it/articles/introduction-systemic-risk-2/.

Collier, Stephen, and Andrew Lakoff. "Vital Systems Security: Reflexive Biopolitics and the Government of Emergency." *Theory, Culture and Society* 32, no. 3 (2015): 19–51.

Collier, Stephen, James Christopher Mizes, and Antina von Schnitzler. "Preface: Public Infrastructures/Infrastructural Publics." *Limn*, no. 7 (2016). https://limn.it/issues/public-infrastructuresinfrastructural-publics/.

Collins, Robert M. *More: The Politics of Economic Growth in Postwar America*. Oxford: Oxford University Press, 2000.

Cooper, Melinda. *Life as Surplus: Biotechnology and Capitalism in the Neoliberal Era*. Seattle: University of Washington Press, 2008.

Cooper, Melinda. "Turbulent Worlds: Financial Markets and Environmental Crisis." *Theory, Culture and Society* 27, no. 2–3 (2010): 167–190.

Cooper, Melinda, and Jeremy Walker. "Genealogies of Resilience: From Systems Ecology to the Political Economy of Crisis Adaptation." *Security Dialogue* 14, no. 2 (2011): 143–160.

Cowen, Deborah. *The Deadly Life of Logistics: Mapping Violence in Global Trade*. Minneapolis: University of Minnesota Press, 2014.

Crist, Meghan. "Postcards from the Edge of Consciousness." *Nautilus*, no. 27 (August 6, 2015). http://nautil.us/issue/27/dark-matter/postcards-from-the-edge-of-consciousness-rp.

Crocq, Marc-Antoine, and Louis Crocq. "From Shell Shock and War Neurosis to Posttraumatic Stress Disorder: A History of Psychotraumatology." *Dialogues in Clinical Neuroscience* 2, no. 1 (2000): 47–55.

Crosby, J. L. "Computers in the Study of Evolution." *Science Progress* 55 (1967): 279–229.

Crowther-Heyck, Hunter. *Economics, Bounded Rationality and the Cognitive Revolution.* Brookfield, VT: Edward Elgar, 1992.

Crowther-Heyck, Hunter. *Herbert A. Simon: The Bounds of Reason in Modern America.* Baltimore: Johns Hopkins University Press, 2005.

Darwin, Charles. *The Origin of Species by Means of Natural Selection; or the Preservation of Favoured Races in the Struggle for Life.* London: John Murray, 1859.

Das, Satyajit. *Traders, Guns, and Money: Knowns and Unknowns in the Dazzling World of Derivatives.* Edinburgh: Prentice Hall, 2006.

Daston, Lorraine. *Classical Probability in the Enlightenment.* Princeton, NJ: Princeton University Press, 1995.

Daston, Lorraine. "The Rule of Rules, or How Reason Became Rationality." Lecture, Max Planck Institute for the History of Science, November 21, 2010, Kindle.

Datta, Ayona. "Postcolonial Urban Futures: Imagining and Governing India's Smart Urban Age." *Environment and Planning D: Society and Space* 37, no. 3 (2019): 393–410.

Davidson, Justin. "MoMA's Schemes for Fixing Urban Problems Are Either Too Dainty or Too Sweeping." *Vulture*, November 20, 2014. https://www.vulture.com /2014/11/review-momas-uneven-growth.html.

DeGrove, J. M. *Planning Policy and Politics: Smart Growth and the States.* Cambridge, MA: Lincoln Institute of Land Policy, 2005.

Dehaene, Stanislas. *Reading in the Brain: The New Science of How We Read.* New York: Viking, 2009.

Deleuze, Gilles. "Postscript on the Societies of Control." *October* 59 (1992): 3–7.

DeLoughrey, Elizabeth M. "The Myth of Isolates: Ecosystem Ecologies in the Nuclear Pacific." *Cultural Geographies* 20, no. 2 (2012): 167–184.

Desrosières, Alain. *The Politics of Large Numbers: A History of Statistical Reasoning.* Translated by Camille Naish. Cambridge, MA: Harvard University Press, 1998.

Dogru, Tarik, Makarand Mody, Courtney Suess, Nathan Line, and Mark Bonn. "Airbnb 2.0: Is It a Sharing Economy Platform or a Lodging Corporation?" *Tourism Management* 78 (2020): 1–5.

Dreier, Peter. "Downtown Development and Urban Reform: The Politics of Boston's Linkage Policy." *Urban Affairs Quarterly* 26, no. 3 (1991): 354–370.

Du, Ding-Zhu, Panos M. Pardalos, and Weili Wu. "History of Optimization." In *Encyclopedia of Optimization*, edited by Christodoulos A. Floudas and Panos M. Pardalos. New York: Springer, 2008, 1538–1542.

Dumit, Joseph. "Circuits in the Mind." Lecture, New York University, 2007, 1–27.

Dumit, Joseph. *Drugs for Life: How Pharmaceutical Companies Define Our Health.* Durham, NC: Duke University Press, 2012.

Easterling, Keller. *Extrastatecraft: The Power of Infrastructure Space.* New York: Verso, 2014.

Edwards, Paul N. *The Closed World: Computers and the Politics of Discourse in Cold War America.* Cambridge, MA: MIT Press, 1997.

Edwards, Paul N. *A Vast Machine: Computer Models, Climate Data, and the Politics of Global Warming.* Cambridge, MA: MIT Press, 2010.

Einhorn, Catrin, and Christopher Flavelle. "A Race against Time to Rescue a Reef from Climate Change." *New York Times,* December 5, 2020. https://www.nytimes.com/2020/12/05/climate/Mexico-reef-climate-change.html.

Elichirigoity, Fernando. *Planet Management: Limits to Growth, Computer Simulation, and the Emergence of Global Spaces.* Evanston, IL: Northwestern University Press, 1999.

Ellwood, David W. *Rebuilding Europe: Western Europe, America, and Postwar Reconstruction.* London: Longman, 1992.

Energy Information Administration, US Department of Energy. *Derivatives and Risk Management in the Petroleum, Natural Gas, and Electricity Industries.* October 2002. http://econometricainc.com/wp-content/uploads/2016/08/EIA_Derivatives_Report.pdf.

Erickson, Paul, Judy L. Klein, Lorraine Daston, Rebecca M. Lemov, Thomas Sturm, and Michael D. Gordin. *How Reason Almost Lost Its Mind: The Strange Career of Cold War Rationality.* Chicago: University of Chicago Press, 2015.

Escobar, Arturo. *Designs for the Pluriverse: Radical Interdependence, Autonomy, and the Making of Worlds.* Durham, NC: Duke University Press, 2018.

Escobar, Arturo. *Encountering Development: The Making and Unmaking of the Third World.* 2nd ed. Princeton, NJ: Princeton University Press, 2012.

Ewald, François. *The Birth of Solidarity: The History of the French Welfare State.* Durham, NC: Duke University Press, 2020.

Fejerskov, Adam Moe. "The New Technopolitics of Development and the Global South as a Laboratory of Technological Experimentation." *Science, Technology, and Human Values* 42 (2017): 947–968.

Filotas, Elise, Lael Parrott, Philip J. Burton, Robin L. Chazdon, K. David Coates, Lluís Coll, Sybille Haeussler, et al. "Viewing Forests through the Lens of Complex Systems Science." *Ecosphere* 5, no. 1 (2014): 1–23.

Fishman, Robert. *Urban Utopias in the Twentieth Century: Ebenezer Howard, Frank Lloyd Wright, and Le Corbusier.* Cambridge, MA: MIT Press, 1982.

Flavelle, Christopher. "A Climate Plan in Texas Focuses on Minorities. Not Everyone Likes It." *New York Times*, July 24, 2020. https://www.nytimes.com/2020/07/24 /climate/houston-flooding-race.html.

Fogel, David B., ed. *Evolutionary Computation: The Fossil Record.* Hoboken, NJ: John Wiley and Sons, 1998.

Fogel, David B. "An Introduction to Simulated Evolutionary Optimization." *IEEE Transactions of Neural Networks* 5, no. 1 (1994): 3.

Fogel, L. I., A. J. Owens, and M. I. Walsh. "Artificial Intelligence through a Simulation of Evolution." In *Biophysics and Cybernetic Systems: Proceedings of the 2nd Cybernetic Sciences Symposium,* edited by M. Maxfield, A. Callahan, and L. J. Fogel, 131–155. Washington, DC: Spartan Books, 1965.

Foley, Duncan. "The Strange History of the Economic Agent." Paper presented to the General Seminar of the Graduate Faculty, New School University, New York, December 6, 2002. https://www.researchgate.net/publication/228563317_The_strange_history_of _the_economic_agent#:~:text=The%20curious%20history%20of%20the,spread%20 of%20capitalist%20social%20relations.&text=It%20turns%20out%2C%20as%20 well,such%20as%20physics%20and%20biology.

Folkers, Andreas. "Continuity and Catastrophe: Business Continuity Management and the Security of Financial Operations." *Economy and Society* 46, no. 1 (2017): 103–127.

Folkers, Andreas. *Das Sicherheitsdispositiv der Resilienz: Katastrophische Risiken und die Biopolitik vitaler Systeme.* Frankfurt: Campus, 2018.

Forrester, Jay W. *Industrial Dynamics.* Cambridge, MA: MIT Press, 1961.

Forrester, Jay W. *Urban Dynamics.* Cambridge, MA: MIT Press, 1969.

Foucault, Michel. *The Birth of Biopolitics: Lectures at the Collège de France, 1978–79.* Edited by M. Senellart. Translated by G. Burchell. New York: Palgrave Macmillan, 2008.

Foucault, Michel. *The History of Sexuality,* vol. 1, *An Introduction.* Translated by Robert Hurley. New York: Pantheon Books, 1978.

Foucault, Michel. *Security, Territory, Population: Lectures at the Collège de France, 1977–78.* Edited by M. Senellart. Translated by G. Burchell. New York: Palgrave Macmillan, 2007.

Foucault, Michel. *Society Must Be Defended: Lectures at the Collège de France, 1975–76.* Edited by Mauro Bertani and Alessandro Fontana. Translated by D. Macey. New York: Picador, 2003.

Furuhata, Yuriko. *Climatic Media: Transpacific Experiments in Atmospheric Control.* Durham, NC: Duke University Press, 2022.

Fraser, A. S. "Simulation of Genetic Systems by Automatic Digital Systems. I. Introduction." *Australian Journal of Biological Sciences* 10 (1957): 484–491.

Friedman, Milton. "The Methodology of Positive Economics." In *Essays in Positive Economics,* 3–43. Chicago: University of Chicago Press, 1953.

Friedman, Milton. "The Need for Futures Markets in Currencies (December 20, 1971)." *Cato Journal* 31, no. 3 (2011): 635–641.

Furuhata, Yuriko. "Multimedia Environments and Security Operations: Expo '70 as a Laboratory of Governance." *Grey Room*, no. 54 (Winter 2014): 56–79.

Gabrys, Jennifer. *Program Earth: Environmental Sensing Technology and the Making of a Computational Planet*. Minneapolis: University of Minnesota Press, 2016.

Galison, Peter. *Image and Logic*. Chicago: University of Chicago Press, 1997.

Galison, Peter. "The Ontology of the Enemy: Norbert Wiener and the Cybernetic Vision." *Critical Inquiry* 21 (1994): 228–266.

Galloway, Alexander R. *Protocol: How Control Exists after Decentralization*. Cambridge, MA: MIT Press, 2004.

Gallun, Alby, and Micah Maidenberg. "Reckless Abandon: The Foreclosure Crisis Is Ruining Swaths of Chicago—Will It Pull Down the Entire City?" *Crain's: Chicago Business*, November 9, 2013.

Garcia-Likens, Patty. "NTUA, SRP and Navajo Nation Leaders Celebrate Ground-breaking Ceremony for Kayenta II." SRP, August 27, 2018, https://media.srpnet.com /kayenta-ii-groundbreaking-ceremony/.

Géron, Aurélien. *Hands-On Machine Learning with Scikit-Learn and TensorFlow: Concepts, Tools, and Techniques to Build Intelligent Systems*. Sebastopol, CA: O'Reilly Media, 2017.

Gerrits, Lasse, and Peter Marks. "The Evolution of Wright's (1932) Adaptive Field to Contemporary Interpretations and Uses of Fitness Landscapes in the Social Sciences." *Biology and Philosophy* 30 (2015): 459–479.

Ghosh, Dhrubajyoti. *Ecology and Traditional Wetland Practice: Lessons from Wastewater Utilisation in the East Calcutta Wetlands*. Kolkata: Worldview, 2005.

Glantz, Aaron. *Homewreckers: How a Gang of Wall Street Kingpins, Hedge Fund Magnates, Crooked Banks, and Vulture Capitalists Suckered Millions out of Their Homes and Demolished the American Dream*. New York: HarperCollins, 2019.

Godfrey-Smith, Peter. *Darwinian Populations and Natural Selection*. Oxford: Oxford University Press, 2009.

Goldman Sachs Global Business Continuity Program. *Business Continuity and Technology Resilience Program for Disaster Recovery: Overview*. March 10, 2021. https://www .goldmansachs.com/disclosures/business-continuity.pdf.

Google Ngram Viewer, s.v. "optimization." Accessed November 4, 2016. https://books .google.com/ngrams/graph?content=optimization&year_start=1800&year_end=2000 &corpus=15&smoothing=3&share=&direct_url=t1%3B%2Coptimization%3B%2Cc0.

Government of Canada. "Gold Facts." Accessed July 29, 2020. https://www.nrcan.gc .ca/our-natural-resources/minerals-mining/gold-facts/20514.

Graham, Stephen, and Simon Marvin. *Splintering Urbanism: Networked Infrastructures, Technological Mobilities and the Urban Condition*. London: Routledge, 2001.

Greenfield, Adam. *Against the Smart City (the City Is Here for You to Use)*. New York: Do Projects, 2013.

Greenhalgh, Susan, and Edwin A. Winckler. *Governing China's Population: From Leninist to Neoliberal Biopolitics*. Stanford, CA: Stanford University Press, 2005.

Greenpeace. *It's the Finance Sector, Stupid.* January 21, 2020. https://www.greenpeace.org/static/planet4-international-stateless/2020/01/13e3c75b-greenpeace_report_wef_2020_its-the-finance_sector_stupid.pdf.

Gregg, Melissa. *Counter Productive: Time Management in the Knowledge Economy*. Durham, NC: Duke University Press, 2018.

Gross, Matthias, and Wolfgang Krohn. "Society as Experiment: Sociological Foundations for a Self-Experimental Society." *History of the Human Sciences* 18, no. 2 (2005): 63–86.

Grove, Kevin. *Resilience*. New York: Routledge, 2018.

Grove, Richard H. *Green Imperialism: Colonial Expansion, Tropical Island Edens and the Origins of Environmentalism, 1600–1860*. Cambridge: Cambridge University Press, 1996.

Gunderson, Lance H., Craig R. Allen, and C. S. Holling, eds. *Foundations of Ecological Resilience*. Washington, DC: Island Press, 2009.

Gunderson, Lance H., and C. S. Holling, eds. *Panarchy: Understanding Transformations in Human and Natural Systems*. Washington, DC: Island Press, 2002.

Gúzman, Lorena. "The Fight for the Control of Chile's Lithium Business." *Diálogo Chino*, December 7, 2018. https://dialogochino.net/15614-the-fight-for-control-of-chiles-lithium-business/.

Hacking, Ian. *The Taming of Chance*. Cambridge: Cambridge University Press, 1990.

Hagen, Joel B. *An Entangled Bank: The Origins of Ecosystem Ecology*. New Brunswick, NJ: Rutgers University Press, 1992.

Hale, Piers J. *Political Descent: Malthus, Mutualism, and the Politics of Evolution in Victorian England*. Chicago: University of Chicago Press, 2014.

Halpern, Orit. *Beautiful Data: A History of Vision and Reason since 1945*. Durham, NC: Duke University Press, 2015.

Halpern, Orit. "Cybernetic Rationality." *Distinktion: Scandinavian Journal of Social Theory* 15, no. 2 (2014): 223–238.

Halpern, Orit. "Dreams for Our Perceptual Present: Temporality, Storage, and Interactivity in Cybernetics." *Configurations* 13, no. 2 (2005): 283–319.

Halpern, Orit, and Gokce Günel. "Demoing unto Death: Smart Cities, Environment, and Preemptive Hope." *FibreCulture*, no. 29 (2017). http://twentynine.fibre

culturejournal.org/fcj-215-demoing-unto-death-smart-cities-environment-and
-preemptive-hope/.

Halpern, Orit, Jesse LeCavalier, and Nerea Calvillo. "Test-Bed Urbanism." *Public Culture* 25, no. 2 (March 2013): 272–306.

Hamblin, Jacob Darwin. *Arming Mother Nature: The Birth of Catastrophic Environmentalism.* Oxford: Oxford University Press, 2013.

Hardt, Michael, and Antonio Negri. *Commonwealth.* Cambridge, MA: Belknap Press of Harvard University Press, 2009.

Hardt, Michael, and Antonio Negri. *Empire.* Cambridge, MA: Harvard University Press, 2000.

Hardt, Michael, and Antonio Negri. *Multitude: War and Democracy in the Age of Empire.* New York: Penguin Press, 2004.

Harney, Stefano, and Fred Moten. *The Undercommons: Fugitive Planning and Black Study.* New York: Minor Compositions, 2013.

Harris, Paul. "Chile Seawater Desalination to Grow 156%." *Mining Journal*, January 27, 2020. https://www.mining-journal.com/copper-news/news/1379729/chile-sea water-desalination-to-grow-156.

Harvey, David. *A Brief History of Neoliberalism.* New York: Oxford University Press, 2005.

Harvey, David. *Spaces of Capital.* London: Routledge, 2012.

Hawkins, Mike. *Social Darwinism in European and American Thought, 1860–1945: Nature as Model and Nature as Threat.* Cambridge: Cambridge University Press, 1997.

Hayek, Friedrich. "The Pretence of Knowledge." Lecture, Sveriges Riksbank Prize in Economic Sciences in Memory of Alfred Nobel, December 11, 1974. https://www .nobelprize.org/prizes/economic-sciences/1974/hayek/lecture/.

Hayek, Friedrich. "The Use of Knowledge in Society." *American Economic Review* 35, no. 4 (September 1945): 519–530.

Hayles, N. Katherine. *How We Became Posthuman: Virtual Bodies in Cybernetics, Literature, and Informatics.* Chicago: University of Chicago Press, 1999.

Hebb, Donald. *The Organization of Behavior: A Neuropsychological Theory.* New York: Wiley, 1949.

Heims, Steve. *The Cybernetics Group.* Cambridge, MA: MIT Press, 1991.

Heron, Woodburn. "The Pathology of Boredom." *Scientific American* 196, no. 1 (January 1957): 52–57.

Hharlamov, Rusian, and Heriner Glassbeck. "When Commodities Get Hooked on Derivatives." *Financial Times*, June 13, 2019. https://www.ft.com/content/896e47c8 -8875-11e9-a028-86cea8523dc2.

Hirt, Sonia A. "Split Apart: How Regulations Designated Populations to Different Parts of the City." In *One Hundred Years of Zoning and the Future of Cities*, edited by Amnon Lehavi, 3–26. Cham, Switzerland: Springer, 2018.

History Channel. "Agent Orange." May 16, 2019. https://www.history.com/topics/vietnam-war/agent-orange-1.

Hofstadter, Richard. *Social Darwinism in American Thought*. New York: G. Braziller, 1959.

Holling, C. S., ed. *Adaptive Environmental Assessment and Management*. New York: John Wiley and Sons, 1978.

Holling, C. S. "Engineering Resilience versus Ecological Resilience." In *Engineering within Ecological Constraints*, edited by Peter C. Schulze, 31–44. Washington, DC: National Academy Press, 1996.

Holling, C. S. "Resilience and Stability of Ecological Systems." *Annual Review of Ecological Systems* 4 (1973): 1–23.

Holling, C. S. "Understanding the Complexity of Economic, Ecological, and Social Systems." *Ecosystems* 4, no. 5 (2001): 390–405.

Hookway, Branden. *Pandemonium: The Rise of Predatory Locales in the Postwar World*. Princeton, NJ: Princeton Architectural Press, 1999.

H.R. 9322. Public No. 397. 73d Congress. Act to Provide for the Establishment, Operation, and Maintenance of Foreign-Trade Zones in Ports of Entry of the United States, to Expedite and Encourage Foreign Commerce, and for Other Purposes. 1934.

Hughey, Matthew W. "Of Riots and Racism: Fifty Years since the Best Laid Schemes of the Kerner Commission (1968–2018)." *Sociological Forum* 33, no. 3 (2018): 619–642.

Huque, Sana, Sarmistha Pattanaik, and D. Parthasarathy. "Cityscape Transformation and the Temporal Metamorphosis of East Kolkata Wetlands: A Political Ecology Perspective." *Sociological Bulletin* 69, no. 1 (2020): 95–112.

Hutchinson, G. E. "Circular Causal Systems in Ecology." *Annals of the New York Academy of Sciences* 40 (1948): 221–246.

Jabr, Ferris. "The Social Life of Forests." *New York Times*, December 12, 2020. https://www.nytimes.com/interactive/2020/12/02/magazine/tree-communication-mycorrhiza.html?searchResultPosition=1.

John, Arthur H. "The London Assurance Company and the Marine Insurance Market of the 18th Century." *Economica* 25, no. 98 (1958): 126–141.

Jung, C. G. *The Psychology of Dementia Praecox*. Edited by Frederick W. Peterson and A. A. Brill. New York: Journal of Nervous and Mental Disease, 1909.

Karvonen, Andrew, and Bas Van Heur. "Urban Laboratories: Experiments in Reworking Cities." *International Journal of Urban and Regional Research* 38, no. 2 (2014): 379–392.

Kaufmann, Stuart A. *The Origins of Order: Self-Organization and Selection in Evolution.* Oxford: Oxford University Press, 1993.

Kay, Lily E. "From Logical Neurons to Poetic Embodiments of Mind: Warren McCulloch's Project in Neuroscience." *Science in Context* 14, no. 4 (2001): 591–614.

Keeling, Kara. *Queer Times, Black Futures.* New York: New York University Press, 2019.

Keim, Brandon. "Never Underestimate the Intelligence of Trees." *Nautilus,* no. 77 (October 13, 2019). https://nautil.us/issue/77/underworldsnbsp/never-underestimate -the-intelligence-of-trees.

Kelley, R. D. G. *Freedom Dreams: The Black Radical Imagination.* Boston: Beacon Press, 2002.

King, Melvin. *Chain of Change: Struggles for Black Community Development.* Boston: South End Press, 1981.

King, Nicholas B. "Briefing: Evidence and Uncertainty during the Covid-19 Pandemic." Max Bell School of Public Policy, McGill University, April 6, 2020.https://www .mcgill.ca/maxbellschool/article/briefing-evidence-and-uncertainty-during-covid-19 -pandemic.

Kitchin, Rob. *The Data Revolution: Big Data, Open Data, Data Infrastructures and Their Consequences.* London: Sage, 2014.

Klein, Naomi. "Screen New Deal." *The Intercept,* May 8, 2020. https://theintercept .com/2020/05/08/andrew-cuomo-eric-schmidt-coronavirus-tech-shock-doctrine/.

Klein, Naomi. *The Shock Doctrine: The Rise of Disaster Capitalism.* New York: Metropolitan Books/Henry Holt, 2007.

Knight, Frank. *Risk, Uncertainty, and Profit.* Boston: Houghton Mifflin, 1921.

Kreager, Philip. "Darwin and Lotka: Two Concepts of Population." *Demographic Research* 21, no. 2 (2009): 469–502.

LaDuke, Winona, and Deborah Cowen. "Beyond Wiindigo Infrastructure." *South Atlantic Quarterly* 119, no. 2 (2020): 243–265.

Lanville, Amy N., and Carl D. Meyer. *Google's PageRank and Beyond: The Science of Search Engine Rankings.* Princeton, NJ: Princeton University Press, 2006.

Larkin, Peter. *The Canadian Encyclopedia,* s.v. "Crawford Stanley Holling." Last modified July 11, 2020. https://thecanadianencyclopedia.ca/en/article/crawford-stanley -holling.

Latour, Bruno, and Peter Weibel. *Critical Zones: The Science and Politics of Landing on Earth.* Cambridge, MA: MIT Press, 2020.

Lazzarato, Maurizio. "Immaterial Labour." In *Radical Thought in Italy,* edited by Paolo Virno and Michael Hardt, 132–146. Minneapolis: University of Minnesota Press, 1996.

LeCavalier, Jesse. *The Rule of Logistics: Walmart and the Architecture of Fulfillment.* Minneapolis: University of Minnesota Press, 2016.

Lee, Benjamin, and Edward Lipuma. *Financial Derivatives and the Globalization of Risk.* Durham, NC: Duke University Press, 2004.

Lepore, Jill. *If Then: How the Simulmatics Corporation Invented the Future.* New York: Liveright, 2020.

Lewis, Michael. *Flash Boys: A Wall Street Revolt.* New York: W. W. Norton, 2014.

Light, Jennifer. *From Warfare to Welfare: Defense Intellectuals and Urban Problems in Cold War America.* Baltimore: Johns Hopkins University Press, 2003.

Lindseth, Brian. "The Pre-history of Resilience in Ecological Research." *Limn*, no. 1 (2011). https://limn.it/articles/the-pre-history-of-resilience-in-ecological-research/.

Lovekin, Dave. "Unlocking Clean Energy Opportunities for Indigenous Communities: Federal Funding Will Help Communities Develop Renewable Energy Projects and Transition Off Diesel." Pembina Institute, February 24, 2017. https://www.pembina .org/blog/unlocking-clean-energy-opportunities-indigenous-communities.

Lynch, Kevin. *The Image of the City.* Cambridge, MA: MIT Press, 1960.

MacKenzie, Donald A. *An Engine, Not a Camera: How Financial Models Shape Markets.* Cambridge, MA: MIT Press, 2006.

Malkiel, Burton G., and Eugene F. Fama. "Efficient Capital Markets: A Review of Theory and Empirical Work." *Journal of Finance* 25, no. 2 (1970): 383–417.

Malthus, Thomas. *An Essay on the Principle of Population, as It Affects the Future Improvement of Society. With Remarks on the Speculations of Mr. Godwin, M. Condorcet, and Other Writers.* London: J. Johnson, 1798.

Mareis, Claudia. "Brainstorming Revisited: On Instrumental Creativity and Human Productivity in the Mid-Twentieth Century." *Cultural Politics* 16, no. 1 (2020): 50–69.

Martin, Laura J. "The X-ray Images That Showed Midcentury Scientists How Radiation Affects an Ecosystem." *Slate*, December 28, 2015.

Martin, Randy. "What Difference Do Derivatives Make? From the Technical to the Political Conjuncture." *Culture Unbound* 6 (2014): 189–210.

Masco, Joseph. *The Theater of Operations: National Security Affect from the Cold War to the War on Terror.* Durham, NC: Duke University Press, 2014.

Mattern, Shannon. *Code and Clay, Data and Dirt: Five Thousand Years of Urban Media.* Minneapolis: University of Minnesota Press, 2017.

Mattern, Shannon. *Deep Mapping the Media City.* Minneapolis: University of Minnesota Press, 2015.

Mattern, Shannon. "Interfacing Urban Intelligence." *Places Journal*, April 2014. https:// placesjournal.org/article/interfacing-urban-intelligence/.

Mattern, Shannon. "Methodolatry and the Art of Measure." *Places Journal*, November 2013. https://placesjournal.org/article/methodolatry-and-the-art-of-measure/.

Mattingly, Jonathan. "Declaration of Jonathan Mattingly." Common Cause, et al. v. League of Women Voters of North Carolina, et al. Civil Action No. 1:16-CV-1026-WO-JEP/Civil Action No. 1:16-Cv-1164-WO=JEP. US District Court for the Middle District of North Carolina, March, 6, 2017.

Mayr, Ernst. "Darwin and the Evolutionary Theory in Biology." In *Evolution and Anthropology: A Centennial Appraisal*, edited by B. J. Meggers, 1–10. Washington, DC: Anthropological Society of Washington, 1959.

Mayr, Ernst. *Systematics and the Origin of Species: From the Viewpoint of a Zoologist.* New York: Columbia University Press, 1942.

McCann, Carole R. *Figuring the Population Bomb: Gender and Demography in the Mid-Twentieth Century.* Seattle: University of Washington Press, 2016.

McCulloch, Warren. *Embodiments of Mind.* 1965. Reprint, Cambridge, MA: MIT Press, 1970.

McCulloch, Warren, and Walter Pitts. "A Logical Calculus of Ideas Immanent in Nervous Activity." In *Embodiments of Mind*, edited by Warren McCulloch. 1943. Reprint, Cambridge, MA: MIT Press, 1970.

McGrayne, Sharon Bertsch. *The Theory That Would Not Die: How Bayes' Rule Cracked the Enigma Code, Hunted Down Russian Submarines, and Emerged Triumphant from Two Centuries of Controversy.* New Haven, CT: Yale University Press, 2011.

McKelvey, Fenwick. "The Other Cambridge Analytics: Early 'Artificial Intelligence' in American Political Science." In *The Cultural Life of Machine Learning: An Incursion into Critical AI Studies*, edited by Jonathan Roberge and Michael Castelle, 117–142. Cham, Switzerland: Palgrave Macmillan/Springer Nature Switzerland AG, 2020.

Meadows, Donella H. "The History and Conclusions of *Limits to Growth*." *System Dynamics Review* 23 (2007): 191–199.

Meadows, Donella H., Dennis L. Meadows, Jørgen Randers, and William W. Behrens. *The Limits to Growth: A Report for the Club of Rome's Project on the Predicament of Mankind.* New York: Universe Books, 1972.

Mehrling, Perry. *Fischer Black and the Revolutionary Idea of Finance.* 2005. Reprint, New York: John Wiley and Sons, 2012.

Micale, Mark S., and Paul Lerner, eds. *Traumatic Pasts: History, Psychiatry, and Trauma in the Modern Age, 1870–1930.* Cambridge: Cambridge University Press, 2010.

Ministry of Housing and Urban Affairs, Government of India. Smartnet. Accessed March 9, 2022. https://smartnet.niua.org/.

Ministry of Urban Development, Government of India. *Smart Cities: Mission Statement and Guidelines.* June 2015. http://smartcityrajkot.in/Docs/SmartCityGuidelines.pdf.

Minsky, Marvin. "Steps toward Artificial Intelligence." *Proceedings of the IRE* 49, no. 1 (1961): 8–30.

Mirowski, Philip. "The Future(s) of Open Science." *Social Studies of Science* 48, no. 2 (2018): 171–203.

Mirowski, Philip. "Hell Is Truth Seen Too Late." *Boundary 2* 46, no. 1 (2019): 1–53.

Mirowski, Philip. *Machine Dreams: Economics Becomes a Cyborg Science*. New York: Cambridge University Press, 2002.

Mirowski, Philip. *More Heat than Light: Economics as Social Physics; Physics as Nature's Economics*. Cambridge: Cambridge University Press, 1989.

Mirowski, Philip. *Never Let a Serious Crisis Go to Waste: How Neoliberalism Survived the Financial Meltdown*. New York: Verso, 2014.

Mirowski, Philip. *Science-Mart: Privatizing American Science*. Cambridge, MA: Harvard University Press, 2011.

Mirowski, Philip. "Twelve Theses Concerning the History of Postwar Neoclassical Price Theory." *History of Political Economy* 38 (2006): 343–379.

Mirowski, Philip, and Dieter Plehwe, eds. *The Road from Mont Pèlerin: The Making of the Neoliberal Thought Collective*. Cambridge, MA: Harvard University Press, 2009.

Mitchell, Robert. "Regulating Life: Romanticism, Science, and the Liberal Imagination." *European Romantic Review* 29, no. 3 (2018): 275–293.

Mitchell, Robert. "U.S. Biobanking Strategies and Biomedical Immaterial Labor." *BioSocieties* 7, no. 3 (2012): 224–244.

Mitchell, Robert, and Catherine Waldby. "National Biobanks: Clinical Labour, Risk Production, and the Creation of Biovalue." *Science, Technology, and Human Values* 35, no. 3 (2010): 330–355.

Mitra, Sedeshna. "Roads to New Urban Futures: Flexible Territorialisation in Peri-urban Kolkata and Hyderbad." *Economic and Political Weekly* 53, no. 49 (2018): 56–64.

MIT School of Architecture and Planning. "MIT's Community Fellows Program." Accessed December 10, 2015. http://web.mit.edu/mhking/www/cfp.html.

Mizes, James Christopher. "Who Owns Africa's Infrastructure?" *Limn*, no. 7 (2016). https://limn.it/issues/public-infrastructuresinfrastructural-publics/.

Moorhead, Paul S., and Martin M. Kaplan, eds. *Mathematical Challenges to the Neo-Darwinian Interpretation of Evolution*. Philadelphia: Wistar Institute Press, 1967.

Munn, Luke. "From the Black Atlantic to Black-Scholes: Precursors of Spatial Capitalization." *Cultural Politics* 16, no. 1 (2020): 92–110.

Murphy, Michelle. *Seizing the Means of Reproduction*. Durham, NC: Duke University Press, 2012.

Naimark, Michael. "Aspen the Verb: Musings on Heritage and Virtuality." *Presence Journal* 15, no. 3 (June 2006). http://www.naimark.net/writing/aspen.html.

nArchitects. "Aqueous City." Museum of Modern Art. Accessed November 3, 2016. http://narchitects.com/work/moma-rising-currents/. The original website has disappeared, but the project is available (as of 2021) at https://www.youtube.com/watch?v=O7folAgxX-gaqueous.

National Commission for Truth and Reconciliation. "Report of the Chilean National Commission on Truth and Reconciliation." Santiago, 1990. https://web.archive.org/web/20091124213320/http://www.usip.org/resources/truth-commission-chile-90.

Negroponte, Nicholas. *The Architecture Machine*. Cambridge, MA: MIT Press, 1970.

Negroponte, Nicholas. *Soft Architecture Machines*. Cambridge, MA: MIT Press, 1975.

Negroponte, Nicholas. "Systems of Urban Growth." BA thesis, Massachusetts Institute of Technology, 1965.

Netflix Technology Blog. "Recommending for the World." February 17, 2016. https://netflixtechblog.com/recommending-for-the-world-8da8cbcf051b.

Neutra, Richard. "Comments on Planetary Reconstruction." *Arts and Architecture* 61, no. 12 (December 1944): 20–22.

Neutra, Richard. "Designs for Puerto Rico (a Test Case)." Richard and Dion Neutra Papers, collection no. 1,179. Department of Special Collections, Charles E. Young Research Library, University of California, Los Angeles, 1943.

Neutra, Richard. "Projects of Puerto Rico: Hospitals, Health Centers, and Schools." *Architecture d'Aujourd'hui* 16, no. 5 (May 1946): 71–77.

New York Times. "Coronavirus in the U.S.: Latest Map and Case Count." July 26, 2020. Last modified March 29, 2021. https://www.nytimes.com/interactive/2020/us/coronavirus-us-cases.html.

New York Times. "J. Robert Oppenheimer, Atom Bomb Pioneer, Dies." February 19, 1967. https://archive.nytimes.com/www.nytimes.com/learning/general/onthisday/bday/0422.html.

New York Times. "Resilience." May 3, 2021. https://www.nytimes.com/spotlight/resilience.

Nielsen, Brett, and Ned Rossiter. "The Logistical City." *Transit Labour: Circuits, Regions, Borders*, no. 3 (August 2011): 2–5.

Nik-Khah, Edward. "George Stigler, the Graduate School of Business, and the Pillars of the Chicago School." In *Building Chicago Economics: New Perspectives on the History of America's Most Powerful Economics Program*, edited by Robert Van Horn, Philip Mirowski, and Thomas A. Stapleford, 116–147. Cambridge: Cambridge University Press, 2011.

Nik-Khah, Edward. "The 'Marketplace of Ideas' and the Centrality of Science to Neoliberalism." In *The Routledge Handbook of the Political Economy of Science*, edited by

David Tyfield, Rebecca Lave, Samuel Randalls, and Charles Thorpe, 31–42. London: Routledge, 2017.

Nolan, Ginger. "Quasi-urban Citizenship: The Global Village as 'Nomos of the Modern.'" *Journal of Architecture* 23 (2018): 448–470.

Obama, Barack. *Economic Report of the President, Together with the Annual Report of the Council of Economic Advisers.* January 2017. https://obamawhitehouse.archives.gov /administration/eop/cea/economic-report-of-the-President/2017.

Obama, Barack. The White House, Office of the Press Secretary. *National Security Strategy.* February 6, 2015. https://obamawhitehouse.archives.gov/sites/default/files /docs/2015_national_security_strategy_2.pdf.

O'Bryan, Scott. *The Growth Idea: Purpose and Prosperity in Postwar Japan.* Honolulu: University of Hawai'i Press, 2009.

Ocasio-Cortez, Alexandria. H. Res. 109: Recognizing the Duty of the Federal Government to Create a Green New Deal. 2019.

Odum, Eugene. *Ecology.* New York: Holt, Rinehart, and Winston, 1963.

Odum, Eugene. *Fundamentals of Ecology.* Philadelphia: W. B. Saunders, 1953.

Odum, Howard T. *Systems Ecology: An Introduction.* Edited by Robert L. Metcalf and Werner Stumm. New York: John Wiley and Sons, 1983.

Odum, Howard T., and Eugene P. Odum. "Trophic Structure and Productivity of a Windward Coral Reef Community on Enitwetok Atoll." *Ecological Monographs* 25 (July 1955): 291–320.

Odum, H. T, and R. F. Pigeon, eds. *A Tropical Rain Forest.* Oak Ridge, TN: U.S. Atomic Energy Commission, Division of Technical Information, 1970.

Ong, Aiwha. "Introduction: Worlding Cities, or the Art of Being Global." In *Worlding Cities: Asian Experiments and the Art of Being Global,* edited by Aiwha Ong and Ananya Roy, 1–26. London: Routledge, 2011.

Orff, Kate. "Oyster-tecture." Scape Landscape Architecture DCP. Accessed December 10, 2020. https://www.scapestudio.com/projects/oyster-tecture/.

Overbye, Dennis. "Darkness Visible, Finally: Astronomers Capture First Ever Image of a Black Hole." *New York Times,* April 10, 2019. https://www.nytimes.com/2019/04 /10/science/black-hole-picture.html.

Page, Lawrence, Sergey Brin, Rajeev Motwani, and Terry Winograd. "The PageRank Citation Ranking: Bringing Order to the Web." Technical Report, Stanford InfoLab, 1999. http://ilpubs.stanford.edu:8090/422/.

Palmisano, Sam. "A Smarter Planet: The Next Leadership Agenda." Council on Foreign Relations, November 6, 2008. https://www.youtube.com/watch?v=i_j4-Fm_Svs.

Papanek, Victor. *Design for the Real World: Human Ecology and Social Change.* 2nd ed. Chicago: Academy Chicago Publishers, 1985.

Parikka, Jussi. *Insect Media.* Minneapolis: University of Minnesota Press, 2010.

Park, Robert E. "The City as a Social Laboratory." In *Chicago: An Experiment in Social Science Research,* edited by Thomas Vernor Smith and Leonard Dupee White, 1–19. Chicago: University of Chicago Press, 1929.

Parks, Lisa, and Caren Kaplan. Introduction to *Life in the Age of Drone Warfare,* edited by Lisa Parks and Caren Kaplan, 1–22. Durham, NC: Duke University Press, 2017.

Parks, Lisa, and Janet Walker. "Disaster Media: Bending the Curve of Ecological Disruption and Moving toward Social Justice." *Media + Environment* 2, no. 1 (2020). https://mediaenviron.org/article/13474-disaster-media-bending-the-curve-of-ecological-disruption-and-moving-toward-social-justice.

Pasquinelli, Matteo. "Google's PageRank Algorithm: A Diagram of the Cognitive Capitalism and the Rentier of the Common Intellect." In *Deep Search: The Politics of Search Beyond Google,* edited by Konrad Becker and Felix Stalder, 152–162. London: Transaction Publishers, 2009. http://matteopasquinelli.com/google-pagerank-algorithm/.

Peck, Jamie. *Constructions of Neoliberal Reason.* New York: Oxford University Press, 2010.

Philip, Kavita. "The Internet Will Be Decolonized." In *Your Computer Is on Fire,* edited by Benjamin Peters, Thomas S. Mullaney, Mar Hicks, and Kavita Philip, 91–116. Cambridge, MA: MIT Press, 2021.

Piketty, Thomas. *Capitalism in the 21st Century.* Cambridge, MA: Belknap Press of Harvard University Press, 2014.

Pollio, Andrea. "Uber, Airports, and Labour at the Infrastructural Interfaces of Platform Urbanism." *Geoforum* 118 (January 2021): 47–55.

Poon, Martha. "From New Deal Institutions to Capital Markets: Commercial Consumer Risk Scores and the Making of Subprime Mortgage Finance." *Accounting, Organizations and Society* 34 (2009): 654–674.

Porter, Theodore M. *The Rise of Statistical Thinking, 1820–1900.* Princeton, NJ: Princeton University Press, 1986.

Poundstone, William. *Prisoner's Dilemma.* New York: Doubleday, 1992.

Pratt, Mary Louise. *Imperial Eyes: Travel Writing and Transculturation.* New York: Routledge, 1992.

Rabinbach, Anson. *The Human Motor: Energy, Fatigue, and the Origins of Modernity.* Berkeley: University of California Press, 1992.

Rajan, Kaushik Sunder. *Pharmocracy: Value, Politics, and Knowledge in Global Biomedicine.* Durham, NC: Duke University Press, 2017.

Ramsden, Edmund. "Eugenics from the New Deal to the Great Society: Genetics, Demography and Population Quality." *Studies in History and Philosophy of Biological and Biomedical Sciences* 9 (2008): 391–406.

Ras, Bonnie Riva. "The Navajo Nation's First Solar Project Is Scaling Up." *Goodnet: Gateway to Doing Good*, January 20, 2019. https://www.goodnet.org/articles/navajo-nations-first-solar-project-scaling-up.

Ratti, Carlo, and Matthew Claudel. *The City of Tomorrow: Sensors, Networks, Hackers, and the Future of Urban Life*. New Haven, CT: Yale University Press, 2016.

Rechenberg, I. "Cybernetic Solution Path of an Experimental Problem." *Royal Aircraft Establishment, Library Translations*, no. 1122 (1965).

Rittel, Horst W. J., and Melvin M. Webber. "Dilemmas in a General Theory of Planning." *Policy Sciences* 4 (1973): 155–169.

Robertson, Thomas. "'This Is the American Earth': American Empire, the Cold War, and American Environmentalism." *Diplomatic History* 32, no. 4 (2008): 561–584.

Robinson, Cedric J. *Black Marxism: The Making of the Black Radical Tradition*. 1986. Reprint, Chapel Hill: University of North Carolina Press, 2000.

Roitman, Janet. "Africa Rising: Class or Finance." Podcast, University of Helsinki Seminar, 117 min., April 12, 2020. https://blogs.helsinki.fi/anthropology/2020/12/04/janet-roitman-africa-rising-class-or-finance/.

Rosenblatt, Frank. "The Perceptron: A Probabilistic Model for Information Storage and Organization in the Brain." *Psychological Review* 65, no. 6 (1958): 386–408.

Rosenblatt, Frank. *Principles of Neurodynamics: Perceptrons and the Theory of Brain Mechanisms*. Washington, DC: Spartan Books, 1962.

Rossiter, Ned. *Software, Infrastructure, Labor: A Media Theory of Logistical Nightmares*. New York: Routledge, 2016.

Rothstein, W. G. *Public Health and the Risk Factor: A History of an Uneven Medical Revolution*. Rochester, NY: University of Rochester Press, 2003.

Royal Society of Chemistry Periodic Table. "Lithium." Accessed March 15, 2021. https://www.rsc.org/periodic-table/element/3/lithium.

Sandbach, Francis. "The Rise and Fall of the Limits to Growth Debate." *Social Studies of Science* 8, no. 4 (1978): 495–520.

Sassen, Saskia. *Expulsions: Brutality and Complexity in the Global Economy*. Cambridge, MA: Harvard University Press, 2014.

Schermerhorn, Calvin. *The Business of Slavery and the Rise of American Capitalism, 1815–1860*. New Haven, CT: Yale University Press, 2015.

Schmelzer, Matthias. *The Hegemony of Growth: The OECD and the Making of the Economic Growth Paradigm*. Cambridge: Cambridge University Press, 2016.

Schmitt, Carl. *The Crisis of Parliamentary Democracy*. Translated by Ellen Kennedy. Cambridge, MA: MIT Press, 1985.

Schuerman, Matthew. "Climate Change Fears Meet Development at the New Hudson Yards." WYNC News, December 12, 2012. https://www.wnyc.org/story/256763-under side-developing-hudson-yards/.

Scott, Felicity. "Aspen Proving Grounds." Paper presented at Technics and Art: Architecture, Cities, and History after Mumford, Columbia University, 2012.

Selfridge, Oliver G. "Pandemonium: A Paradigm for Learning." In *Proceedings of the Symposium on Mechanisation of Thought Processes*, edited by D. V. Blake and A. M. Uttley, 511–529. London: Her Majesty's Stationery Office, 1959.

Sennett, Richard. "The Stupefying Smart City." Paper presented at the Urban Age Electric City Conference, London School of Economics, December 2012.

Shah, Nishant. "Identity and Identification: The Individual in the Time of Networked Governance." *Socio-Legal Review* 11, no. 2 (2015): 22–40.

Shannon, Claude, and Warren Weaver. *The Mathematical Theory of Communication*. 1963. Reprint, Urbana-Champaign: University of Illinois Press, 1998.

Shaw, Matt. "Why Arata Isozaki Deserves the Pritzker." *The Architect's Newspaper*, April 16, 2019. https://www.archpaper.com/2019/04/arata-isozaki-pritzker/.

Sidewalk Toronto. "Quayside: A Complete Community and a Proving Ground for Innovation." Accessed December 7, 2020. https://www.sidewalktoronto.ca/plans /quayside/.

Simard, Suzanne W. "Interspecific Carbon Transfer in Ectomycorrhizal Tree Species Mixtures." PhD diss., Oregon State University, 1995.

Simard, Suzanne W. "Nature's Internet: How Trees Talk to Each Other in a Healthy Forest." TEDxSeattle, February 2, 2017. https://www.youtube.com/watch?v=breDQqrkikM.

Simard, Suzanne W., David A. Perry, Melanie D. Jones, David D. Myrold, Daniel M. Durall, and Randy Molina. "Net Transfer of Carbon between Ectomycorrhizal Tree Species in the Field." *Nature* 388 (August 7, 1997): 579–582.

Simon, Dan. *Evolutionary Optimization Algorithms: Biologically-Inspired and Population-Based Approaches to Computer Intelligence*. Hoboken, NJ: John Wiley and Sons, 2013.

Simon, Herbert A. "A Behavioral Model of Rational Choice." *Quarterly Journal of Economics* 69, no. 1 (1955): 99–118.

Simon, Herbert A. "Information-Processing Explanations of Understanding." In *The Nature of Thought: Essays in Honor of D. O. Hebb*, edited by Peter W. Jusczyk and Raymond M. Klein. Hillsdale, NJ: Lawrence Erlbaum, 1980.

Simon, Herbert A. "Rational Decision Making." *American Economic Review* 69, no. 4 (1979): 493–513.

Simon, Herbert A. *The Sciences of the Artificial*. Cambridge, MA: MIT Press, 1996.

Simon, Herbert A., Massimo Egidi, and Robin Marris. *Economics, Bounded Rationality and the Cognitive Revolution*. Aldershot, UK: Edward Elgar, 1995.

Slayton, Rebecca. "Efficient, Secure, Green: Digital Utopianism and the Challenge of Making the Electrical Grid 'Smart.'" *Information and Culture* 48, no. 4 (2013): 448–478.

Spigel, Lynn. "Designing the Smart House: Posthuman Domesticity and Conspicuous Production." In *Electronic Elsewheres: Media, Technology, and the Experience of Social Space*, edited by Chris Berry, Soyoung Kim, and Lynn Spigel, 55–92. Minneapolis: University of Minnesota Press, 2009.

Srnicek, Nick. *Platform Capitalism*. New York: Polity Press, 2016.

Staff reporter. "Codelco to Deploy AI Solution." *Mining Journal*, March 26, 2019. https://www.mining-journal.com/innovation/news/1359598/codelco-to-deploy-ai-solution.

Starosielski, Nicole. *Media Hot and Cold*. Durham, NC: Duke University Press, 2021.

Steenson, Molly Wright. *Architectural Intelligence*. Cambridge, MA: MIT Press, 2017.

Steenson, Molly Wright. "Architectures of Information: Christopher Alexander, Cedric Price, and Nicholas Negroponte and the MIT Architecture Machine Group." PhD diss., Princeton University, 2014.

Sterk, Tristan d'Estrée. "Building upon Negroponte: A Hybridized Model of Control Suitable for Responsive Architecture." *Automation in Construction* 14, no. 2 (2014): 225–232.

Stockholm Resilience Centre. "About Us." Stockholm University, 2020. https://www.stockholmresilience.org/about-us.html.

Strauss, Leo. *Natural Right and History*. Chicago: University of Chicago Press, 1953.

Sundararjan, Arun. *The Sharing Economy: The End of Employment and the Rise of Crowd-Based Capitalism*. Cambridge, MA: MIT Press, 2016.

Sverdrup, H. U., Martin W. Johnson, and Richard H. Fleming. *The Oceans: Their Physics, Chemistry, and General Biology*. New York: Prentice Hall, 1942.

Swiss Re. "Designing a New Type of Insurance to Protect the Coral Reefs, Economies and the Planet." News release, December 10, 2019. https://www.swissre.com/our-business/public-sector-solutions/thought-leadership/new-type-of-insurance-to-protect-coral-reefs-economies.html.

Swiss Re. "Protecting Coral Reefs against Hurricane Damage." *2018 Corporate Responsibility Report*, 2018. https://reports.swissre.com/corporate-responsibility-report/2018/cr-report/solutions/strengthening-risk-resilience-2018-highlights/protecting-coral-reefs-against-hurricane-damage.html.

Szpiro, George G. *Pricing the Future: Finance, Physics, and the 300 Year Journey to the Black-Scholes Equation*. New York: Basic Books, 2011. Kindle.

Taleb, Nassim. *Antifragile: Things That Gain from Disorder.* New York: Random House, 2012.

Tavares, Frank, ed. "Cooking up the World's Driest Desert—Atacama Rover Astrobiology Drilling Studies." NASA, June 20, 2018. https://www.nasa.gov/image-feature /ames/cooking-up-the-world-s-driest-desert-atacama-rover-astrobiology-drilling -studies.

Taylor, Peter J. "Technocratic Optimism, H.T. Odum, and the Partial Transformation of Ecological Metaphor after World War II." *Journal of the History of Biology* 21 (Summer 1988): 213–244.

Taylor, Peter J. *Unruly Complexity: Ecology, Interpretation, Engagement.* Chicago: University of Chicago Press, 2010.

Tkacz, Nathaniel. *Wikipedia and the Politics of Openness.* Chicago: University of Chicago Press, 2014.

Townsend, Anthony M. *Smart Cities: Big Data, Civic Hackers, and the Quest for a New Utopia.* New York: W. W. Norton, 2014.

Turner, Fred. *The Democratic Surround: Multimedia and American Liberalism from World War II to the Psychedelic Sixties.* Chicago: University of Chicago Press, 2013.

van Doorn, Niels. "A New Institution on the Block: On Platform Urbanism and Airbnb Citizenship." *New Media and Society* 22, no. 10 (2020): 1808–1826.

Virilio, Paul. *Speed and Politics.* Cambridge, MA: MIT Press, 2006.

Virno, Paolo. *A Grammar of the Multitude: For an Analysis of Contemporary Forms of Life.* New York: Semiotext(e), 2004.

Vogl, Joseph. "Taming Time: Media of Financialization." *Grey Room* 46 (2012): 72–83.

von Neumann, John. "The General and Logical Theory of Automata." In *Papers of John von Neumann on Computing and Computer Theory,* edited by William Aspray, 491–531. Cambridge, MA: MIT Press, 1986.

von Neumann, John, and Oskar Morgenstern. *Theory of Games and Economic Behavior.* 3rd ed. Princeton, NJ: Princeton University Press, 1990.

Vu, Khoi, Miroslav Begovic, and Damir Novosel. "Grids Get Smart Protection and Control." *IEEE Computer Applications in Power* 10, no. 4 (1997): 40–44.

Wakefield, Stephanie. *Anthropocene Back Loop: Experimentation in Unsafe Operating Space.* London: Open Humanities Press, 2020. http://openhumanitiespress.org/books /download/Wakefield_2020_Anthropocene-Back-Loop.pdf.

Wakefield, Stephanie, and Bruce Braun. "Governing the Resilient City." *Environment and Planning D: Society and Space* 32, no. 1 (2014): 4–11.

Wakefield, Stephanie, and Bruce Braun. "Living Infrastructure, Government, and Destitute Power." Unpublished paper, Anthropology of the Anthropocene, Concordia University, October 23, 2015.

Wakefield, Stephanie, and Bruce Braun. "Oystertecture: Infrastructure, Profanation, and the Sacred Figure of the Human." In *Infrastructure, Environment and Life in the Anthropocene*, edited by Kregg Hetherington, 193–215. Durham, NC: Duke University Press, 2019.

Waldby, Catherine, and Robert Mitchell. *Tissue Economies: Blood, Organs, and Cell Lines in Late Capitalism*. Durham, NC: Duke University Press, 2006.

Waldman, Allan R. "OTC Derivatives and Systemic Risk: Innovative Finance or the Dance into the Abyss?" *American University Law Review* 43 (1994): 1023–1090.

Walker, Alissa. "Sidewalk Labs' 'Smart' City Was Destined to Fail." *Curbed*, May 7, 2020. https://archive.curbed.com/2020/5/7/21250678/sidewalk-labs-toronto-smart-city-fail.

Walker, Jeremy, and Melinda Cooper. "Genealogies of Resilience: From Systems Ecology to the Political Economy of Crisis Adaptation." *Security Dialogue* 2 (2001): 143–160.

Wardrip-Fruin, Noah, and Nick Montfort, eds. *The New Media Reader*. Cambridge, MA: MIT Press, 2003.

Warren, Tom. "Microsoft Teams Jumps 70 Percent to 75 Million Daily Active Users." *The Verge*, April 29, 2020. https://www.theverge.com/2020/4/29/21241972/microsoft-teams-75-million-daily-active-users-stats.

Wharton Business Daily. "The Next Silicon Valley? Why Toronto Is a Contender." March 14, 2019. https://knowledge.wharton.upenn.edu/article/the-next-silicon-valley-why-toronto-is-a-contender/.

Wiener, Norbert. *Cybernetics; or, Control and Communication in the Animal and the Machine*. New York: MIT Press, 1961.

Wikipedia, s.v. "Event Horizon." Last modified January 29, 2022. https://en.wikipedia.org/wiki/Event_horizon.

Williams, Frank Backus. *Building Regulation by Districts, the Lesson of Berlin*. New York: National Housing Association, 1914.

Wilson, Mark. "The Story behind 'Flatten the Curve,' the Defining Chart of the Coronavirus." *Fast Company*, March 13, 2020. https://www.fastcompany.com/90476143/the-story-behind-flatten-the-curve-the-defining-chart-of-the-coronavirus.

Wirthman, Lisa. "NYC's Hudson Yards Looks towards the Future of Smart Development." *Forbes*, October 3, 2018. https://www.forbes.com/sites/nuveen/2018/10/03/nycs-hudson-yards-looks-towards-the-future-of-smart-development/?sh=3e3eb79d46b4.

Witt, John Fabian. *The Accidental Republic: Crippled Workingmen, Destitute Widows, and the Remaking of American Law*. Cambridge, MA: Harvard University Press, 2004.

World Inequality Database. "Percentages of 1% National Income Share." https://wid.world/world/#sptinc_p99p100_z/US;FR;DE;CN;ZA;GB;WO/last/eu/k/p/yearly/s/false/5.070499999999999/30/curve/false/country.

Yeager, Kurt E. "Creating the Second Electrical Century." *Public Utilities Fortnightly* 126 (1990): 8.

Zigrand, Jean-Pierre. "Systems and Systemic Risk in Finance and Economics." SRC Special Paper, no. 1, London School of Economics and Political Science, 2014. http://eprints.lse.ac.uk/61220/1/sp-1_0.pdf.

Zuboff, Shoshana. *The Age of Surveillance Capitalism.* New York: Hachette Book Group, 2019.

INDEX